Dust Free Friends   6a architects

AF070222

Dust Free Friends was commissioned by Brussels based Maniera gallery to make furniture that reflected the thinking of our practice, 6a architects, a series of small pieces of domestic furniture that can be made very simply at home, in restricted spaces, with a small number of tools and without specialist skills followed. The lightness and simplicity of the pieces is derived from the way simple plywood furniture is made on construction sites for stools, tables, work benches, steps, changing quickly and without fuss as uses change. The designs also re-examine the long tradition of self-build that has shared the journey through modernism with industry and craft.

In the early 1970's, the Italian designer Enzo Mari published his seminal book of self-build furniture in direct competition with an expanding consumer market. Using rough sawn softwood, held together only by nails, the method matched the adhocism and toolset of the time.

Forty years later, affordable designer consumerism has more or less replaced old fashioned messy DIY from our homes. Apple and Ikea have replaced amateurish tinkering with pleasure and promise sealed inside immaculate capsules.

Today, the new lithium batteries at the heart of our wireless world have also revolutionised what we can do. With the cordless drill and self-tapping screw in one hand and the dust-free precision of the Festool saw in the other, the world of self-made furniture has reopened with unprecedented speed and ease.

In this series, we invite everyone to make his or her own everyday furniture from dressed plywood. The rules are easy to follow, and even more easy to change, to make best use of the sheet of plywood with the smallest number of cuts and least wastage. While almost every household has a cordless drill, rather fewer have a Festool saw. It is a useful addition but this can also be borrowed or rented – try your nearest designer, artist or architect friend, there's bound to be one nearby. Precision and patience are more important than skill. When you've finished please share your work on Dustfreefriends.co.uk.

Materials

The basic material for *Dust Free Friends* is plywood. It is strong in all directions and can receive screws. We recommend using douglas fir plywood for its colour and texture and because no other plywood tightens under a self tapping screw quite so pleasingly. This is the classic, the original America ply. But other types will work too. Avoid the red far-eastern plywood as their surface and edges are rather dull compared with softwood species.

Most plywood sheets are 1.22×2.44m. This is inherited from the imperial 8'×4'. Watch out as some manufacturers produce a metric version that is 1.2×2.4m. Dimensions of *Dust Free Friends* are designed for the former and should be adjusted for the latter.

Dust Free Friends are designed for plywood 19mm and 11mm thick. Douglas fir plywood comes in these thicknesss but other plywood finishes are 18mm and 12mm. Check the thickness of each cutting list before starting.

Tools

To make *Dust Free Friends*, you will need a saw and a cordless drill.

Festool Saw
To cut the plywood, use a Festool circular saw, matching guide and extractor. Festool produce a coordinated set, making extremely precise and dust free cutting easy and fast.

TS55
DFF have been designed for the most common Festool saw, model TS55 EQ. The blade is exactly 2mm thick. The cutting list includes this thickness in the dimensions. If you use another type of saw, check the blade thickness. The traditional Skilsaw often has 3mm blades, but this will not produce dust free friends.

A short extract of Festool Instruction manual TS55 EQ:

Work area safety
1. Keep work area clean and well lit. Cluttered and dark areas invite accidents.
2. Do not operate power tools in explosive atmospheres, such as in the presence of flammable liquids, gases or dust. Power tools create sparks which may ignite the dust or fumes.
3. Keep children and bystanders away while operating a power tool. Distractions can cause you to lose control.

Technical data
Power consumption: 1,200 W
No load speed: 2,000 – 5,200 rpm
Angle of cut: 0° – 45°
Depth of cut at: 90° 55 mm (2.2″)
45° 43mm (1.7″)
Saw blade diameter 160mm (6.3″)
Saw blade hole diameter 20mm (0.79″)

Dust extraction
Particularly for work in closed areas, we recommend that you connect your circular saw to a chip extractor. This will enable you to reduce the dust load in the air, ensuring that your workplace is clean and improving the quality of your work.

Guide system
The guide rails, which are available in different lengths, allow for precise, clean cuts and simultaneously protect the workpiece surface against damage. In conjunction with the extensive range of accessories, exact angled cuts, mitre cuts and fitting work can be completed with the guide system. The option of securing using G-clamps ensures a firm hold and safe working.

Drill, screwdriver and screws
Any cordless drill / driver will do.

The artist Tom Sachs says that one should always pre-drill holes and countersinks for the heads before screwing plywood. Pre-drilling does have the advantage of precision but many modern self-tapping screws can be used without. If you are not used to it, practice first. The art is in releasing the power just as the head is flush with the surface of the plywood.

If the pressure of the screwhead cracks the edge of the plywood surface, revert to Tom Sachs' recommended technique.

Choose screws carefully: brass or nickel finish.

Spax self-tapping screws are recommended for their quality and the appearance of the star head.

For screwing plywood face to face, screws should be 2–5mm shorter than the combined thickness.

For screwing plywood face to end-grain, screws should be between 2 and 3 times the thickness of the face plywood (eg for screwing 19mm plywood onto an edge, use screws 50–60mm long).

Preparation

Dust Free Friends is a series of decorative plywood furniture pieces. The decorative element should always be applied to the full sheet of plywood before cutting so that the colour, pattern or image is enjoyed against the freshly cut natural laminated edges of the plywood.

Wallpaper

Wallpaper offers an infinite range of colour and pattern that add visual depth and playfulness to the plywood. Materially, wallpaper is the first cousin to plywood. As both are industrially processed from the tree to make thin cellulose fibrous sheet materials, they bond together with ease.

Plant or floral patterned wallpapers are the natural friends of plywood and the wallpapers by William Morris in particular resonate with re-awakening the value and pleasure of manual work in the industrial age. Wallpapers from your local DIY can serve just as well.

Most wallpapers are 52cm wide. A sheet of plywood will require 1 roll, using 3 strips. Apply the first strip of wallpaper along the centre of an uncut sheet of plywood. Apply the second and third strips paying special attention to registering the pattern pattern and ensuring the edges are well bonded. Cut off excess, approx. 15cm on each side.

Seals

Ideally *Dust Free Friends* combine the natural surfaces of the plywood face, cut edges, wallpaper and/or paint. However these are porous and fragile. The pieces will either gather dirt or patina, depending on your point of view, particularly on untreated wood. A clear, matt, water based acrylic seal can be pre-applied to the exposed (un-cut) plywood surface. With each coat comes greater protection from stains and hand marks but also encapsulates the material away from touch and the wear and tear of life. Seals should be used with great care to match surface, ageing and use.

Paint

Dust Free Friends can also be made using pre-painted plywood. The lustre of gold and silver paint transforms the surface of plywood without fundamentally changing its hue. The seal must be applied to the plywood before painting gold or silver as if it is applied after it will affect the lustre of the paint.

Apply regular stripes of gold or silver paint by hand across the full sheet of plywood. The natural irregularity of the hand gently contrasts the industrial precision of the material and occasionally matches the surface grain. Pieces made with gold or silver will catch the light in mysterious ways.

Making & Assembly

Cutting
Cut all the components before assembly. Measure twice, cut once. Be precise. No need to erase light pencil measuring lines. Take extreme care not to put fingers anywhere near blade and cut pieces until safety sleeve is lifted and blade automatically cuts out especially with smalll pieces

Edges
External plywood edges should be *eased* with a light pass of fine grade sand paper held on a block at 45 degrees, dragging away from the surface.

Do not over sand. Edges should remain sharp to the eye, but soft to the touch.

Do not sand edges to be connected to other components as they need to be flush and true for a precise joint. Only exposed external edges should be eased.

Assembly
Always assemble furniture on a table. Prepare assembly sequence carefully.

Glue
Dust Free Friends generally do not require glue. However if extra strength is required, then use Gorilla wood glue. Once a joint is assembled with Gorilla glue, do not cut off or sand the natural expansion of the glue outside the joint. This is just another maker's mark.

Corners
Much of the pleasure and practicality of *Dust Free Friends* is in the corners. Some corners are good, some corners are not. Check each design carefully and, if required, cut corners with a surgeon's precision.

Rules
Dust Free Friends are designed according to a few rules. Some are described above while others are implicit in cutting patterns. The pieces are meant to be fun to make and to use. Feel free to make adjustments or simply add to the collection.

Catalogue

		Dimensions (mm)	Plywood thickness	Page	Sheets needed
9.0 screen		1500×242	11mm	12	1
10.0 step & stool / column		405×270×360 405×270×190	19mm	16	1
11.1 coffee table		600×600×300	11mm	20	1
11.2 coffee table		800×800×300	19mm	24	1
11.3 stool / side table		400×400×400	19mm	28	1
11.4 coffee table		1045×605×329	19mm	32	1
11.5 day bed		2055×673×329	19mm	34	2
12.0 coffee table		1033×533×323	19mm	36	1

13.1 coffee table		600×600×309	11mm	40	1
13.2 stool / side table		400×400×417	19mm	44	1
13.3 nested tables		≤ 500×500×500	11mm	48	1
13.4 console		900×300×811	11mm	50	1
14.0 writing table		1065×609×700	11mm	54	1
15.1 corner shelf		400×400×1220	11mm	58	offcuts from 9.0 screen
15.2 picture rail		size variable	11 and 19mm	62	offcuts
15.3 nail box		391×200×250	19mm	64	offcuts

9.0 screen 11mm

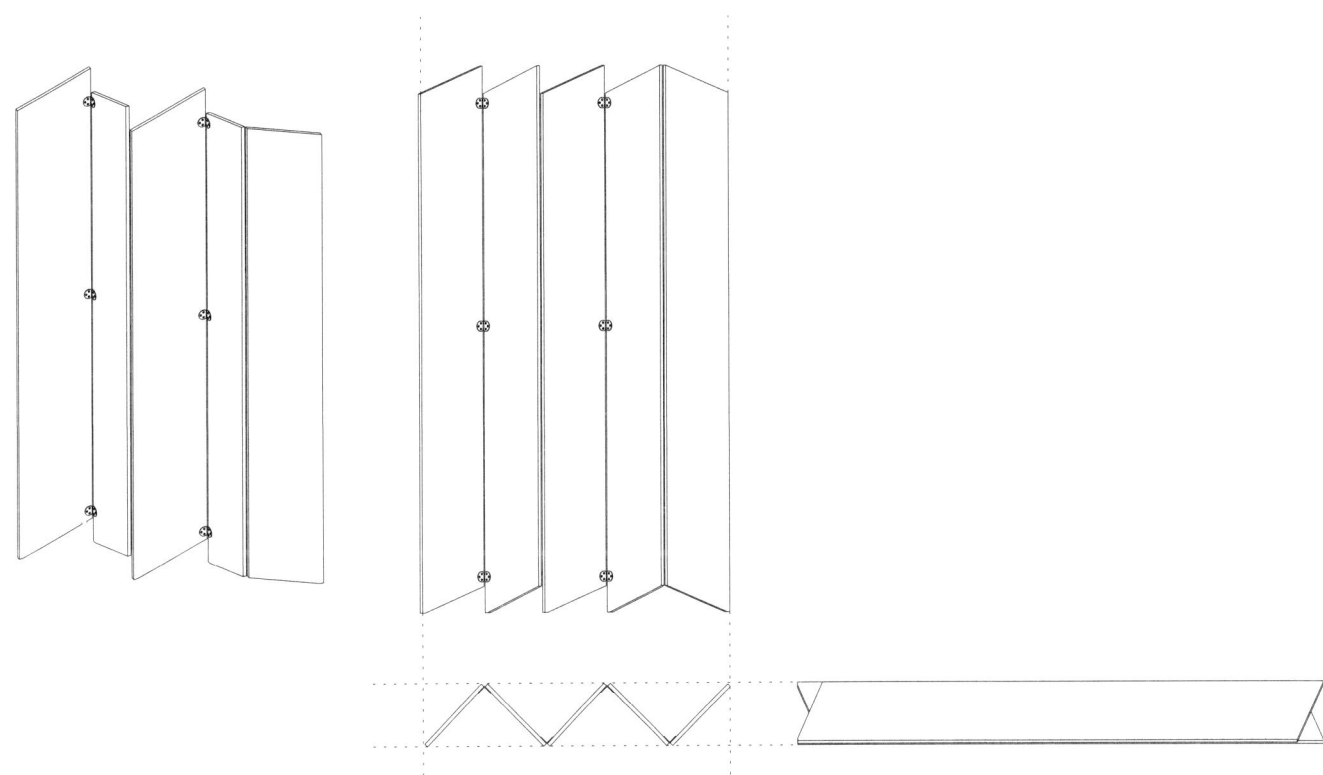

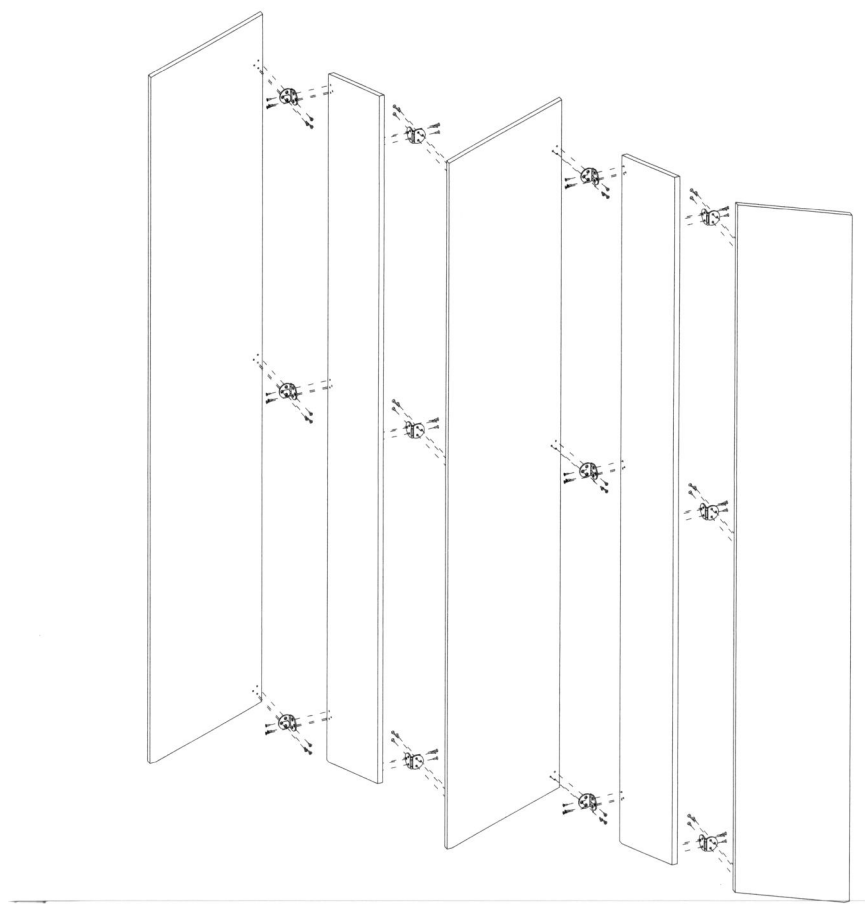

10.0 step & stool / column   19mm

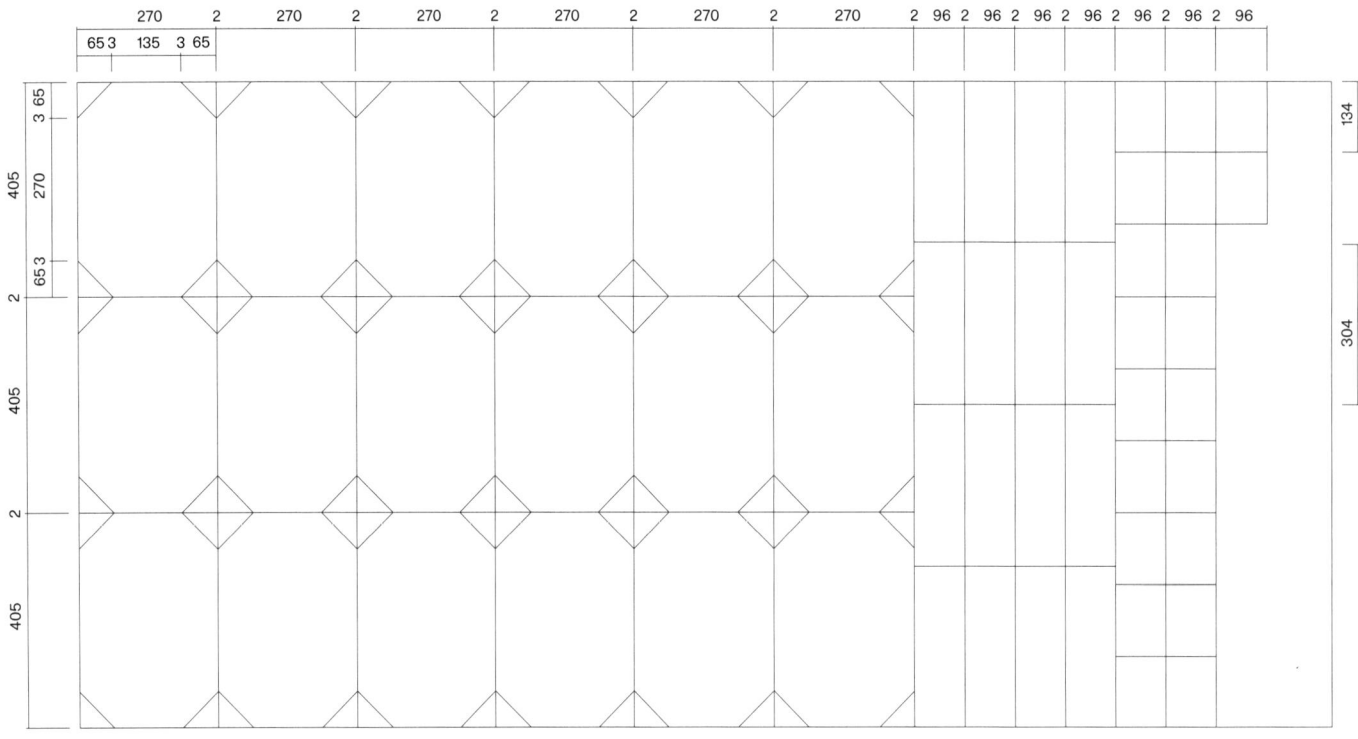

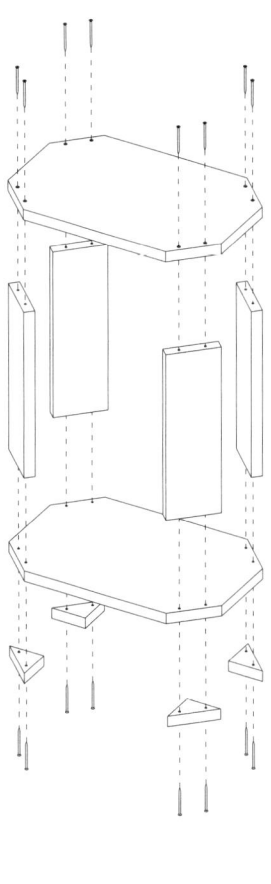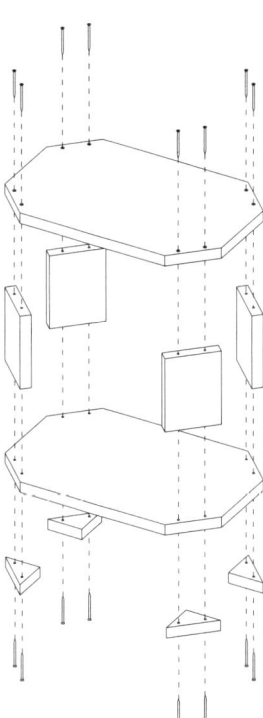

11.1 coffee table   11mm

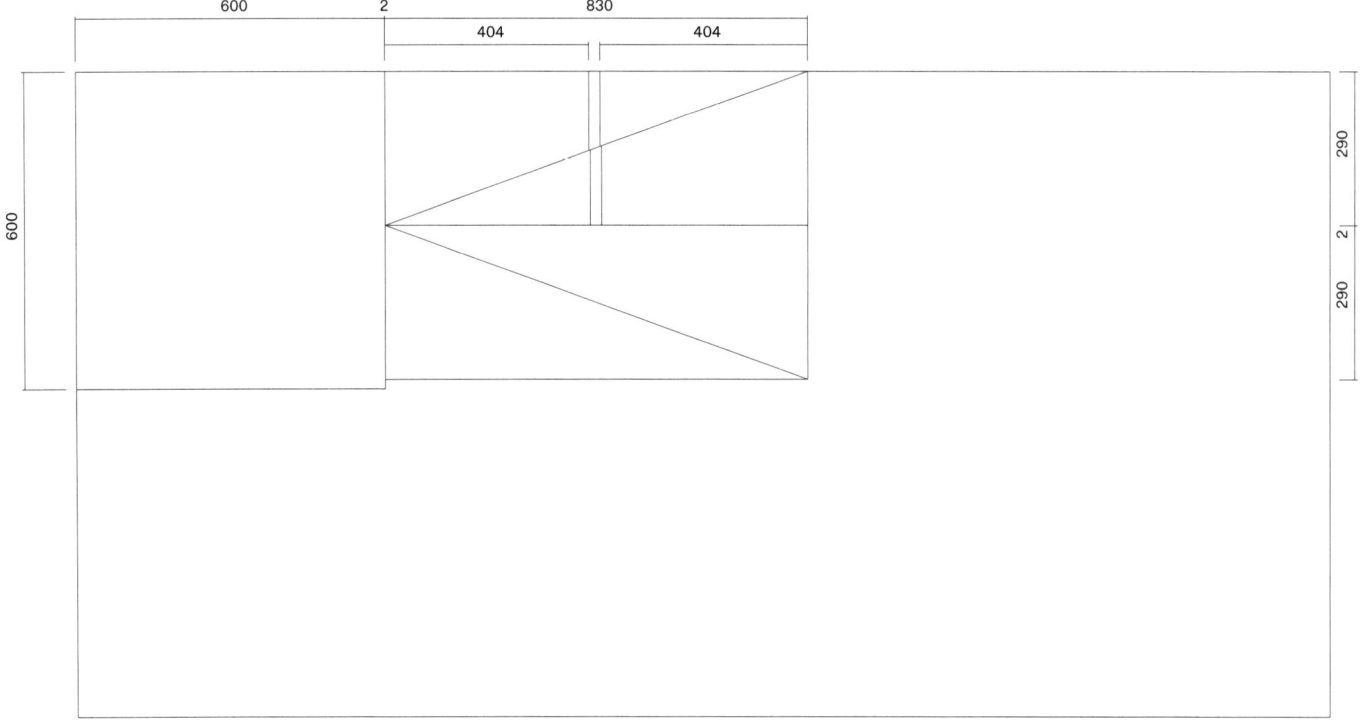

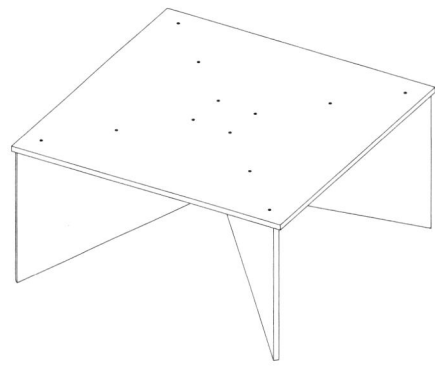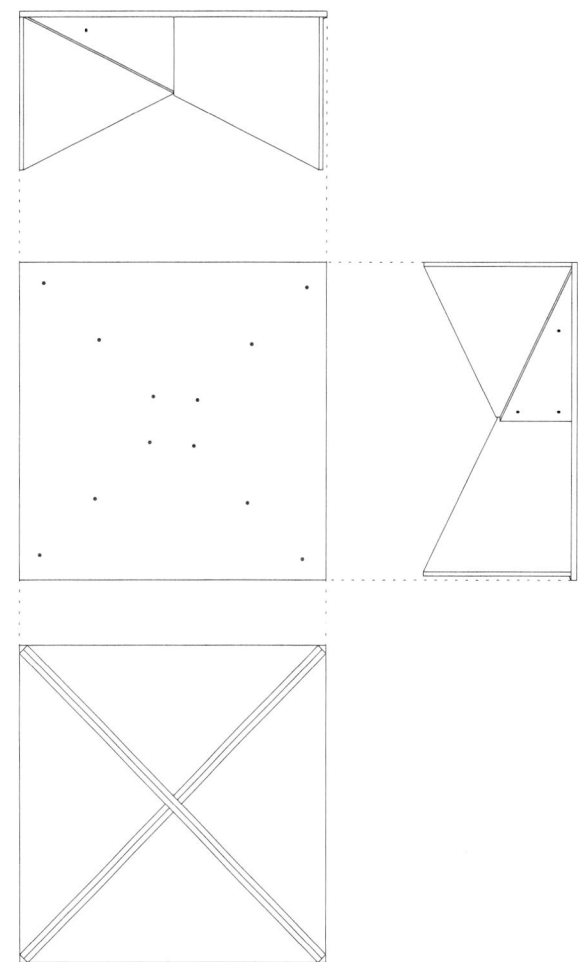

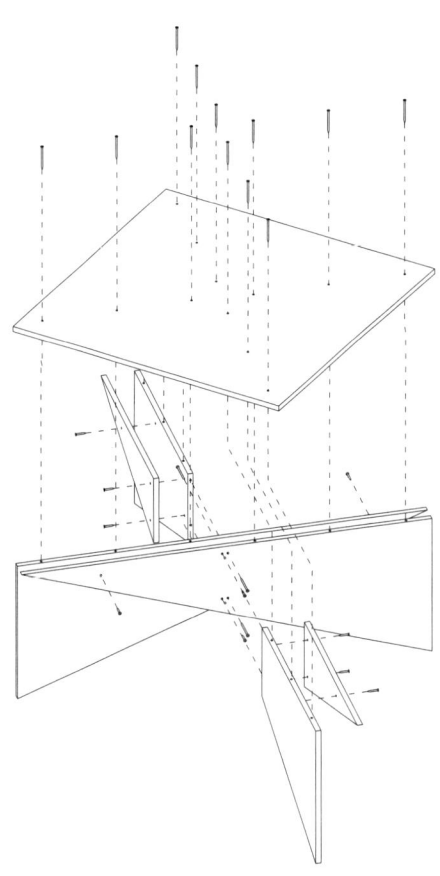

11.2 coffee table   19mm

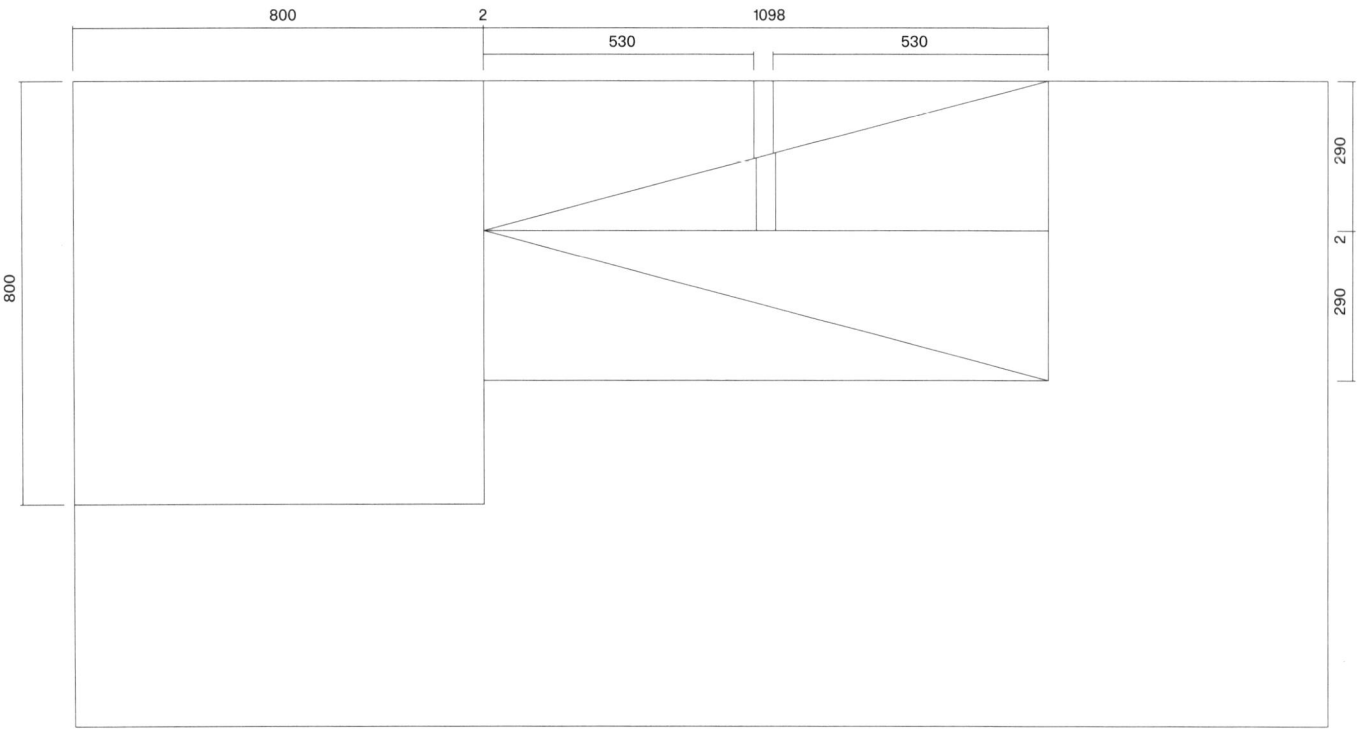

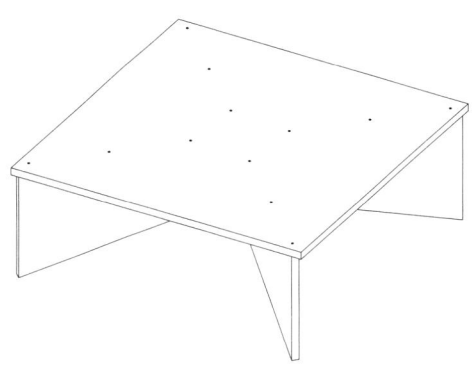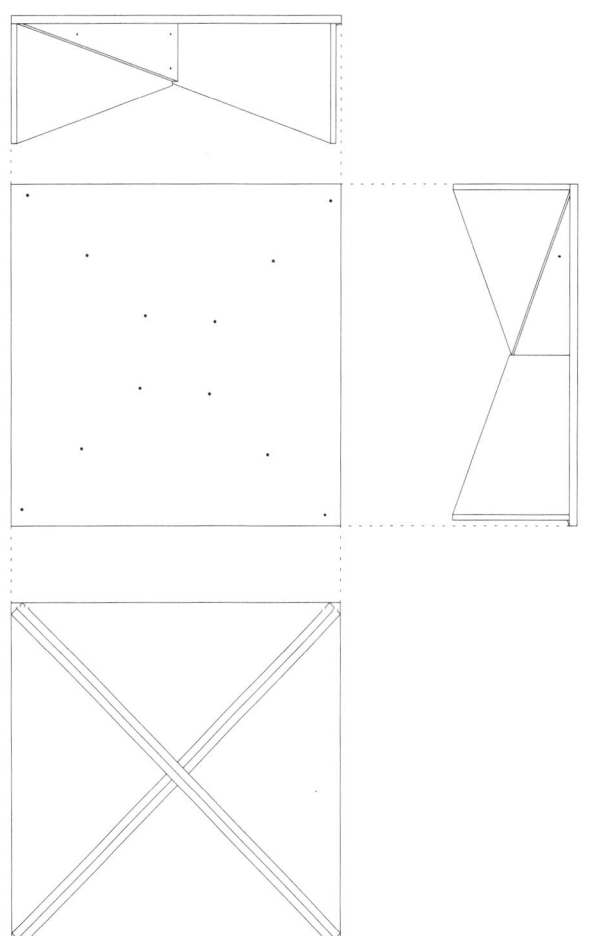

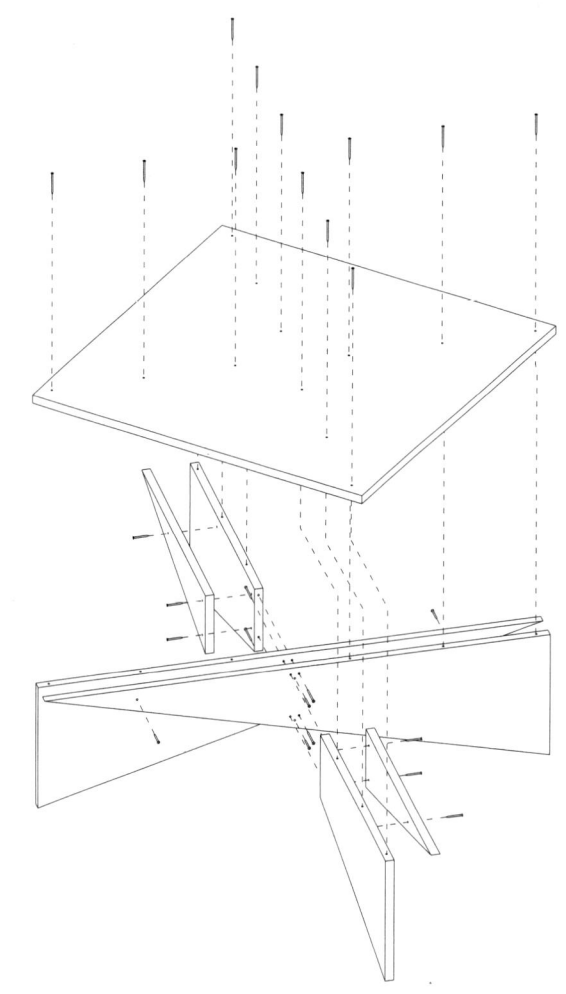

11.3 stool / side table   19mm

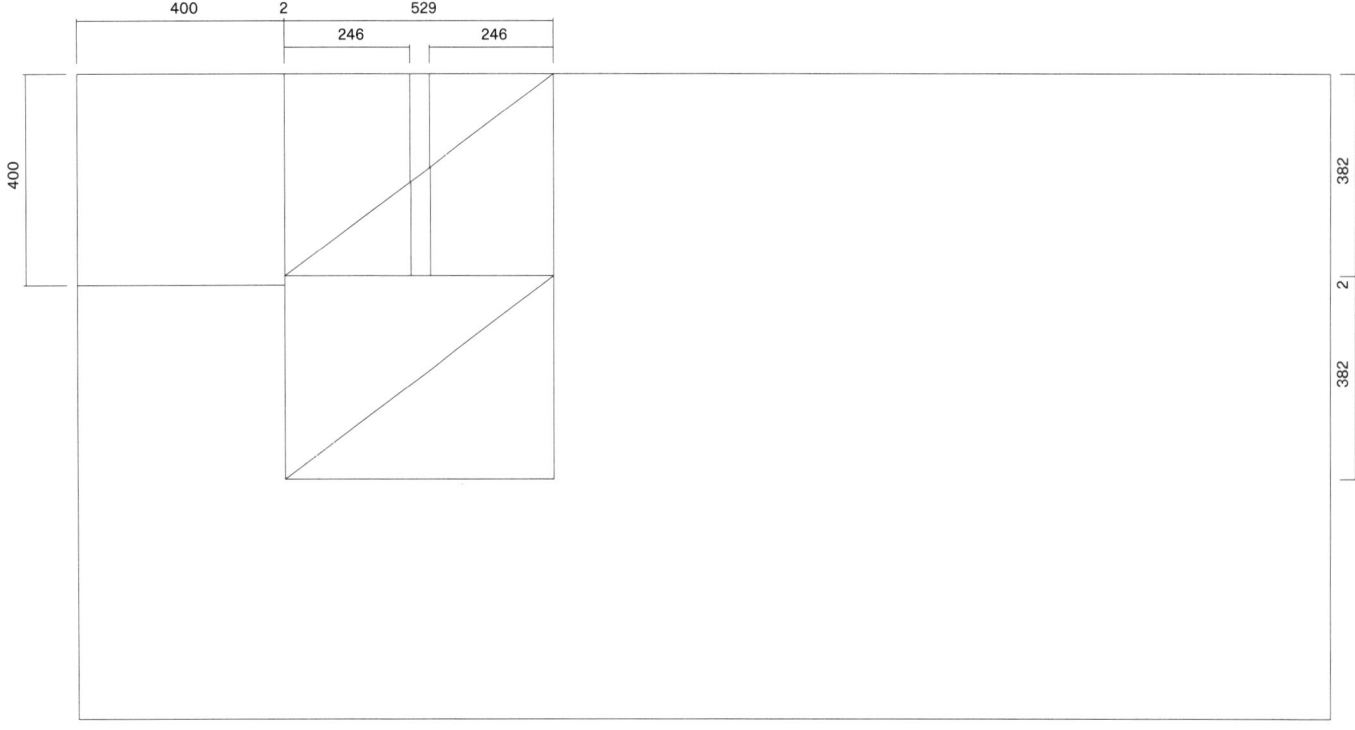

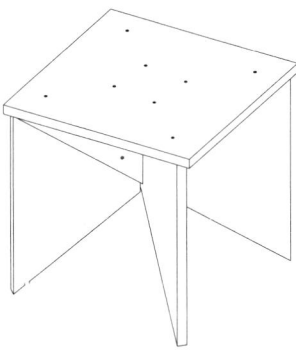
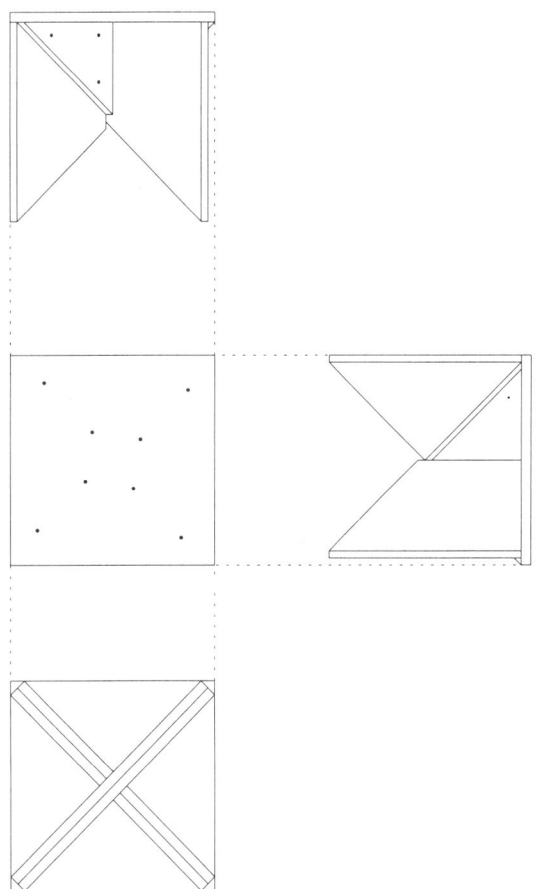

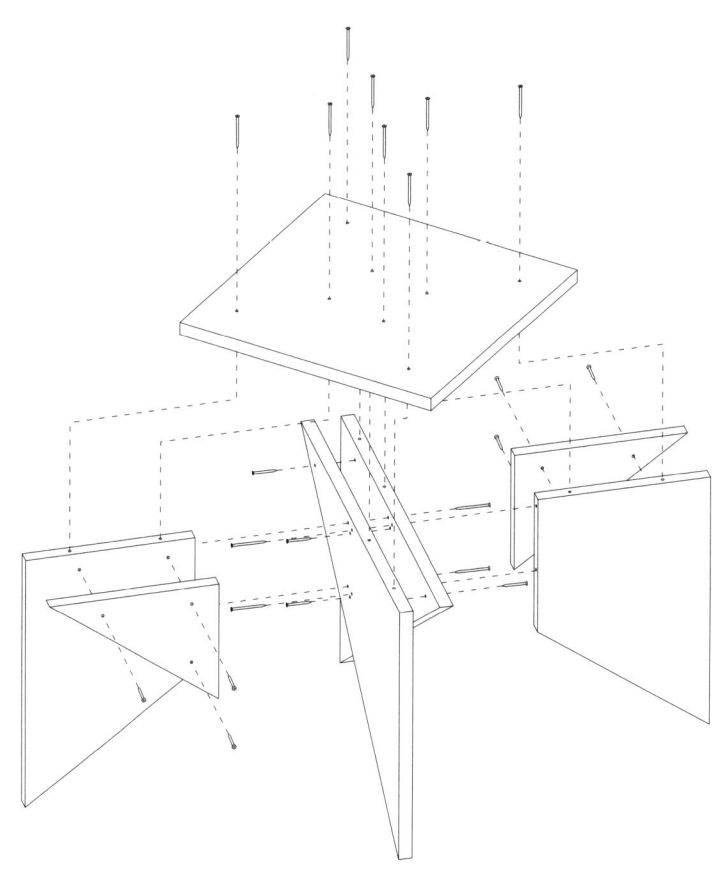

11.4 coffee table 19mm

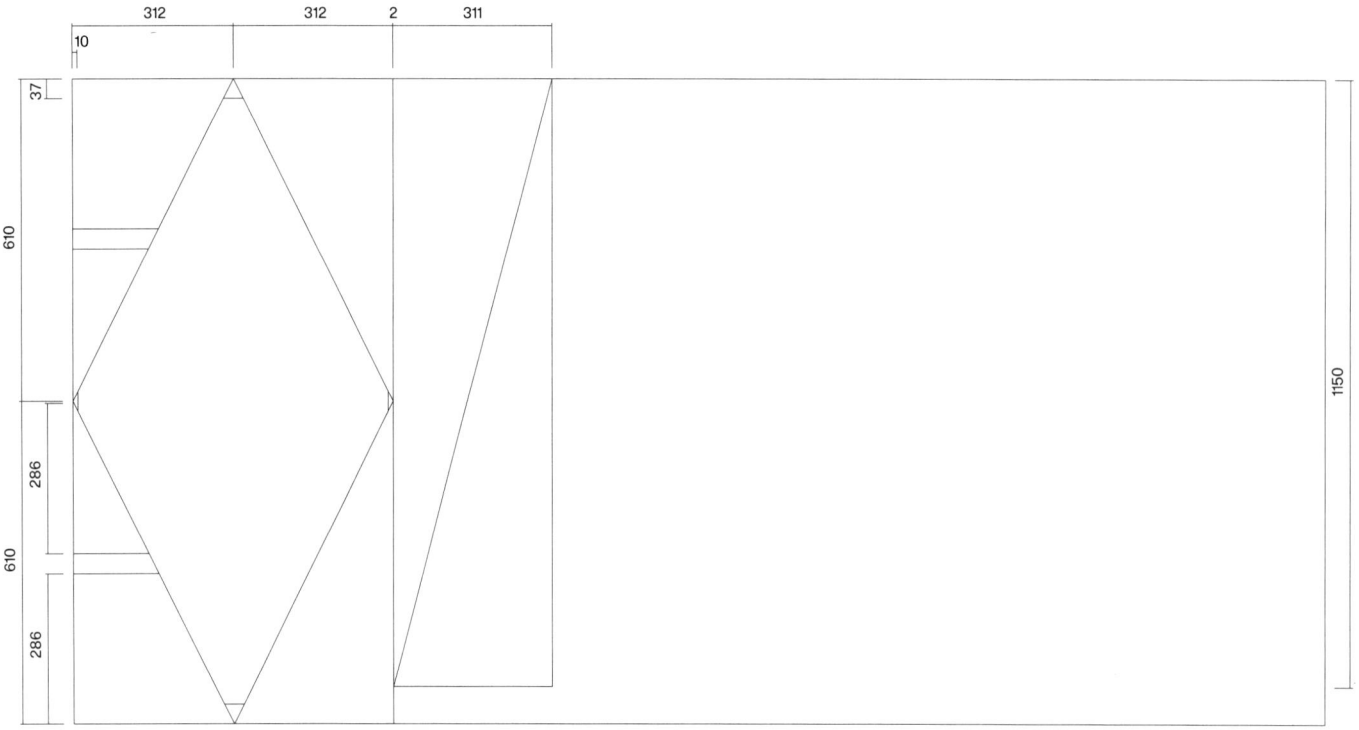

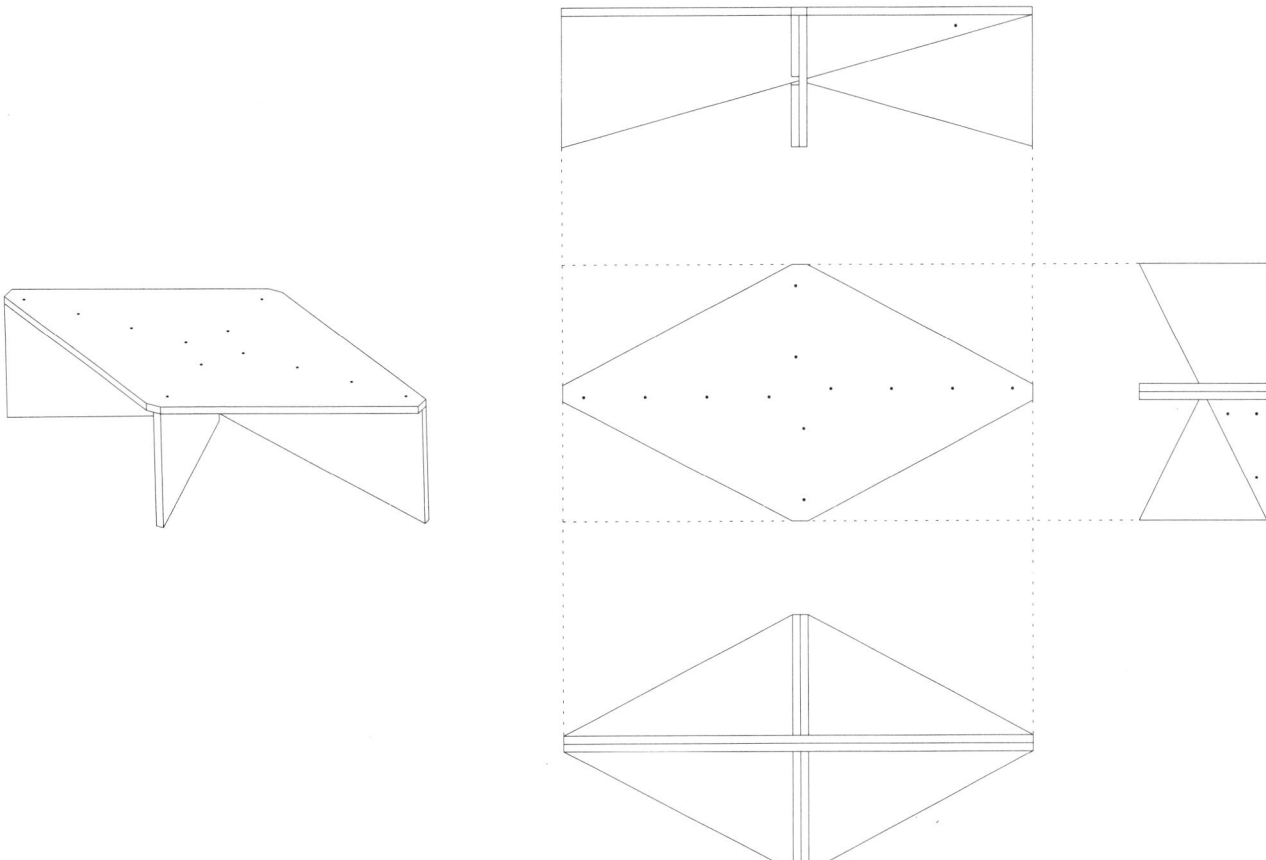

11.5 day bed   19mm

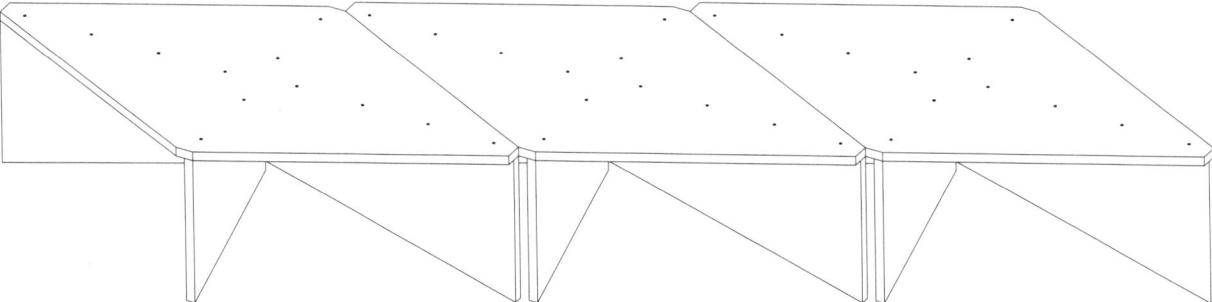

12.0 coffee table   19mm

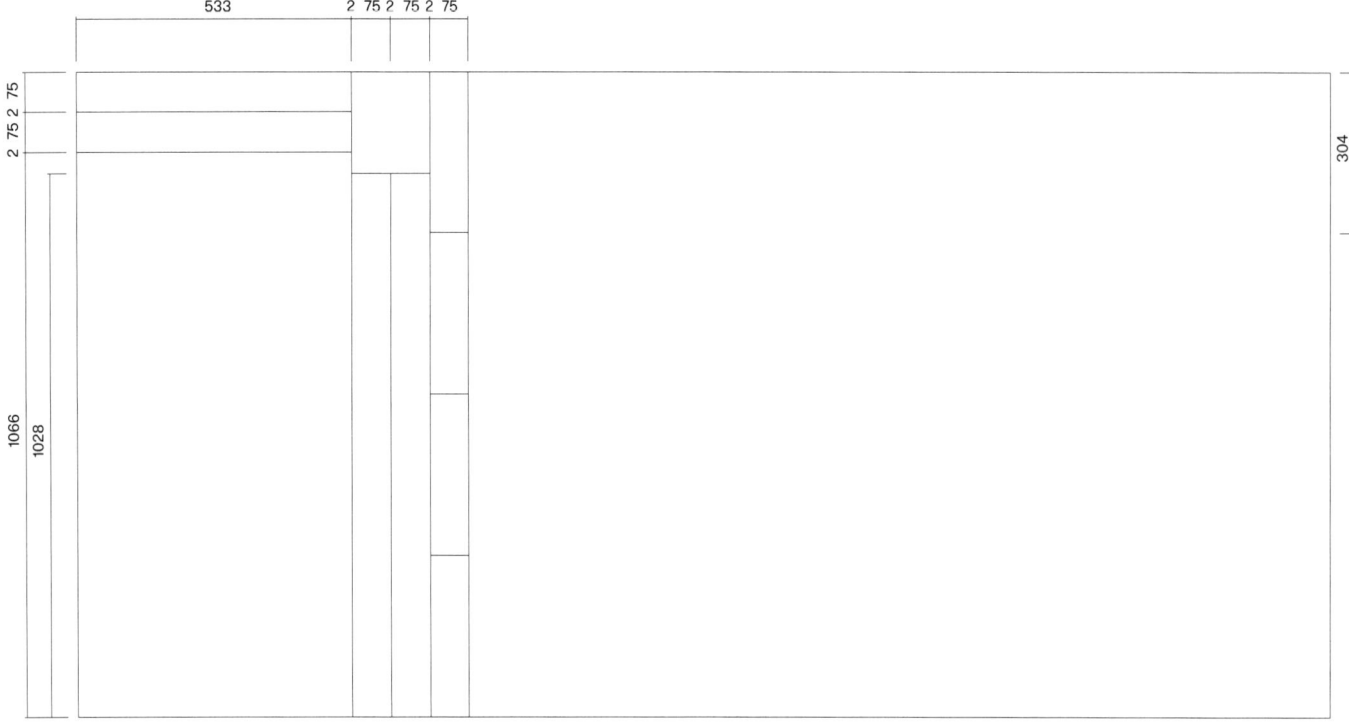

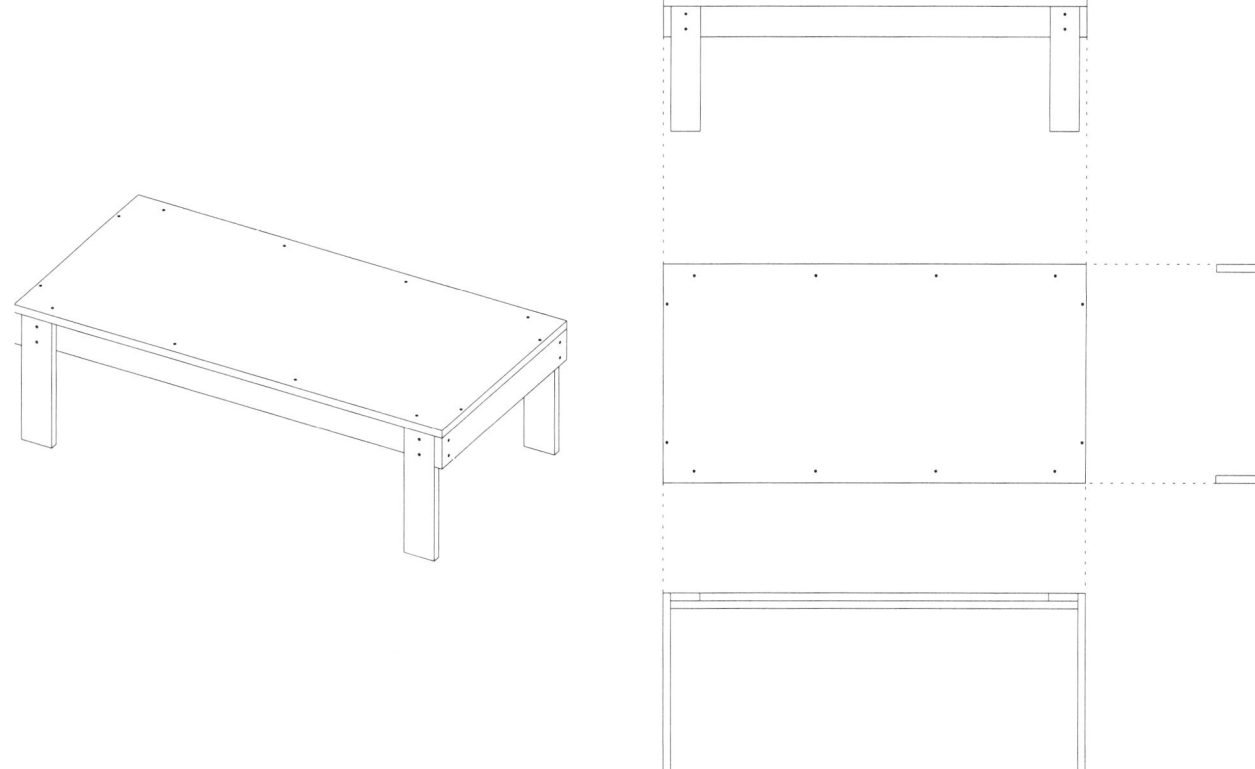

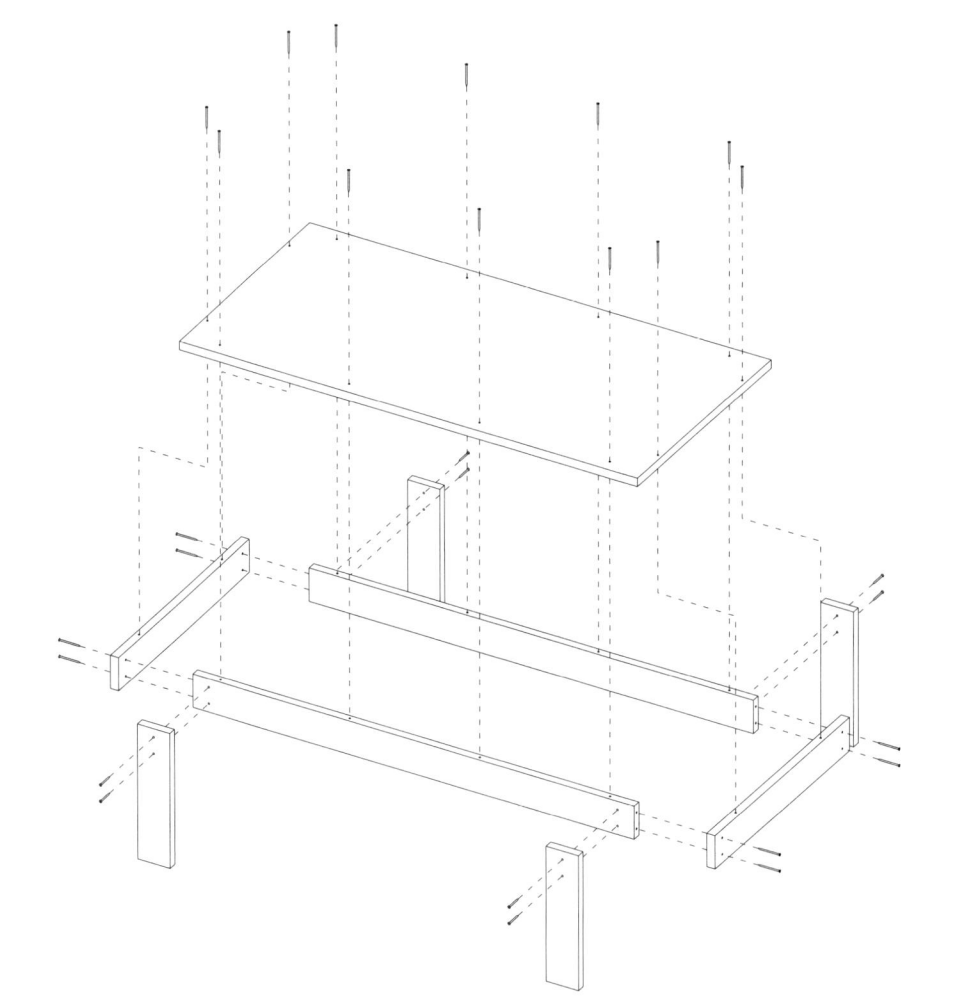

13.1 coffee table 11mm

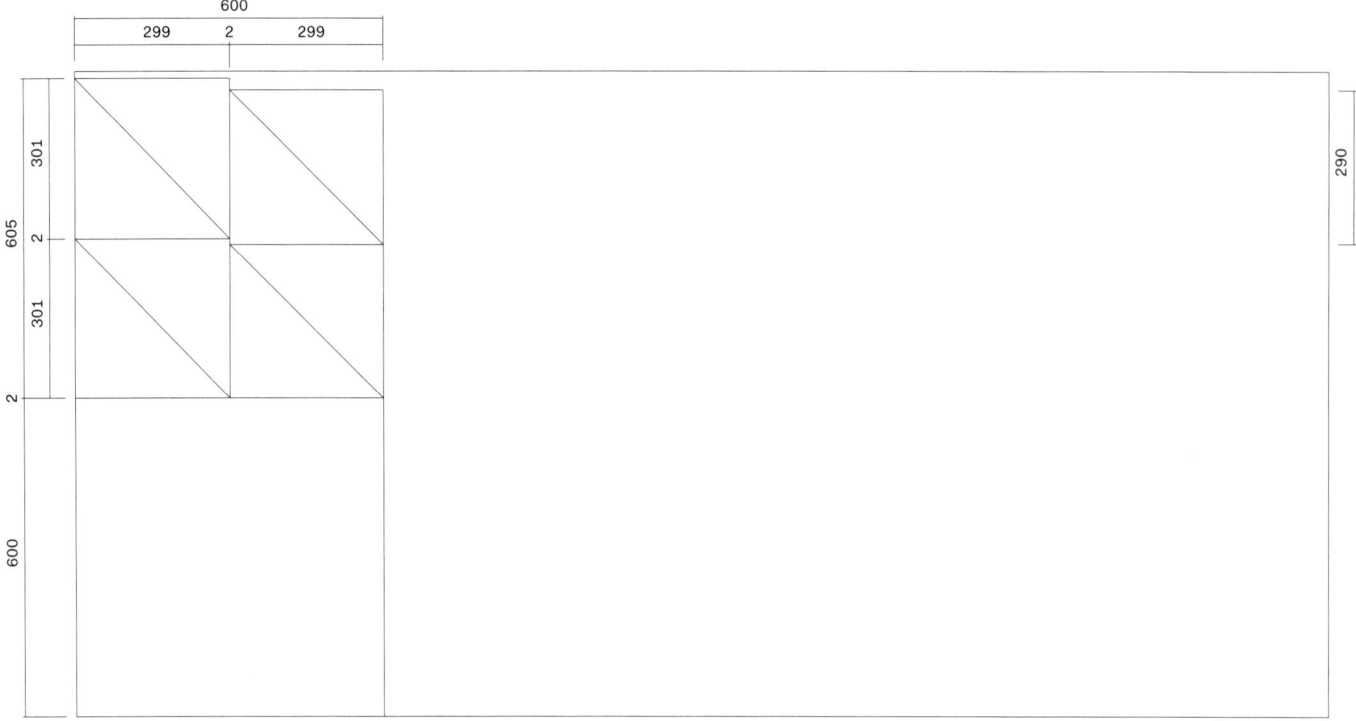

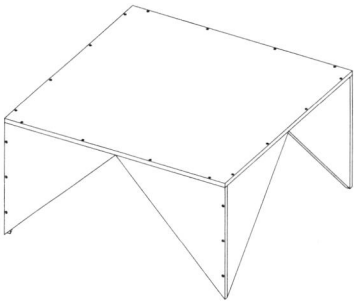
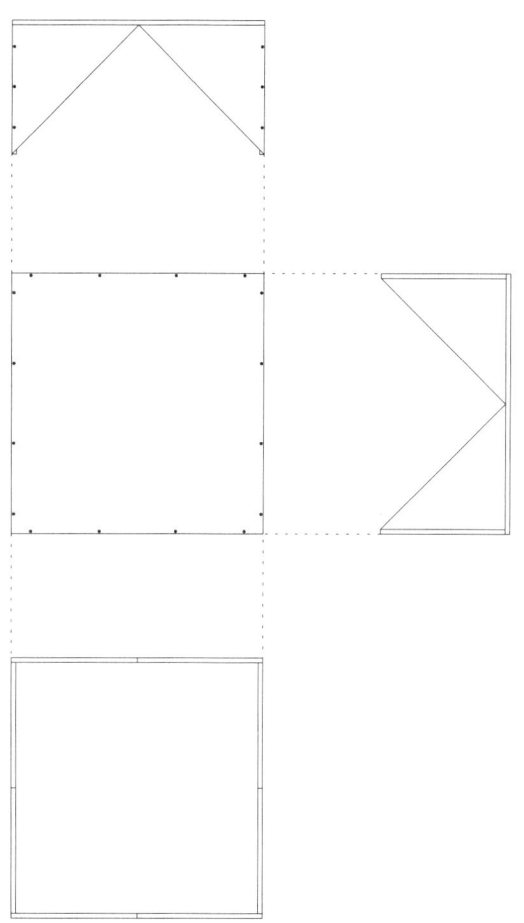

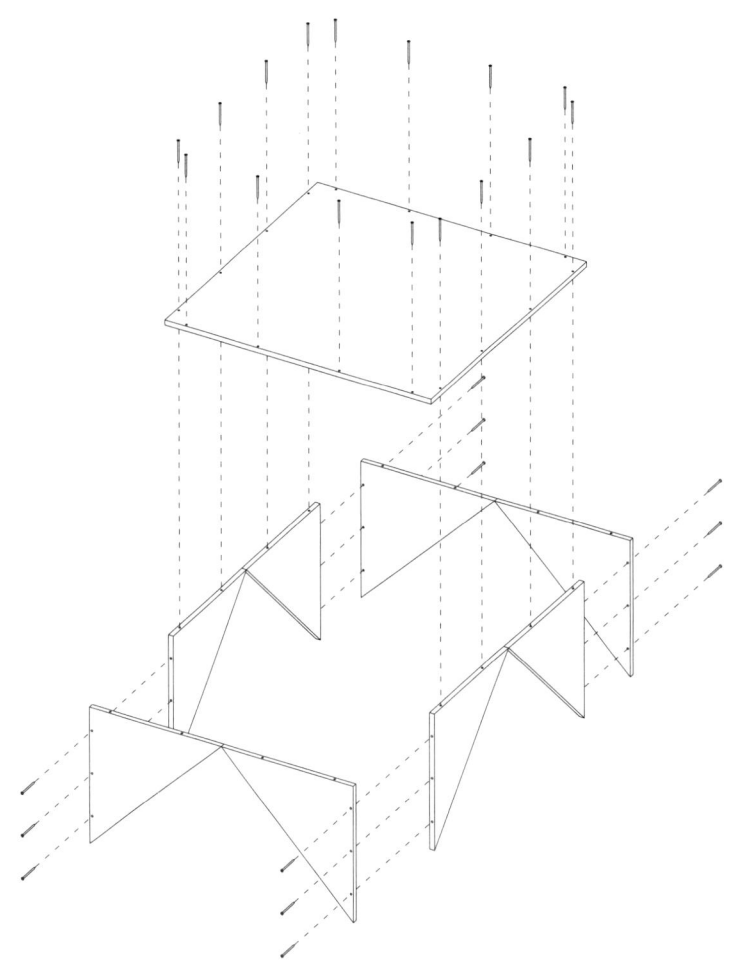

13.2 stool / side table 19mm

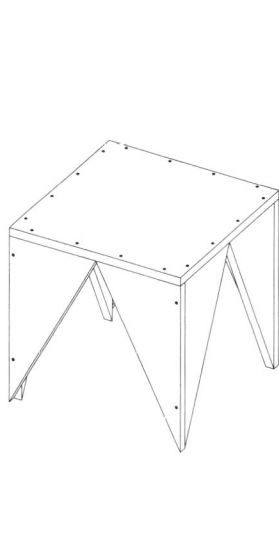
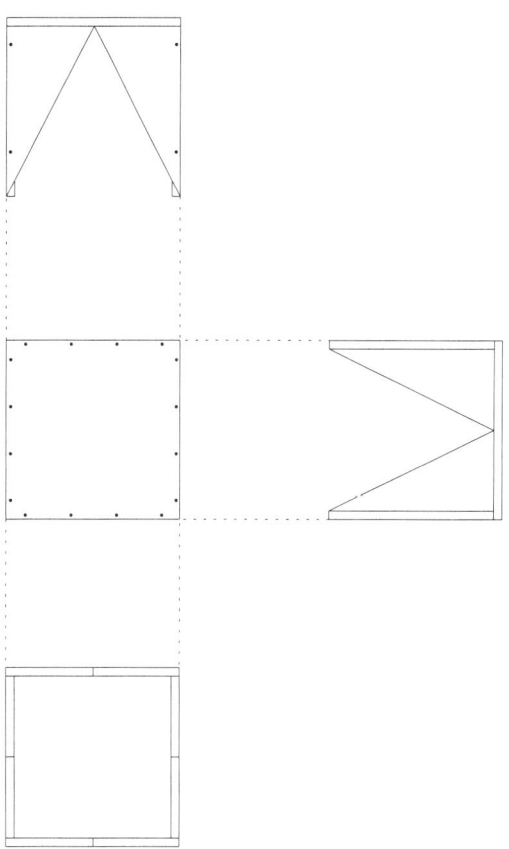

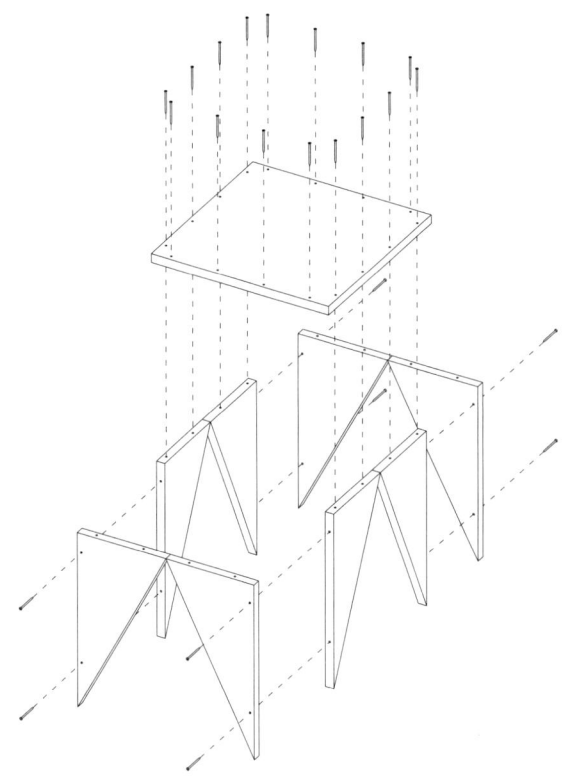

13.3 nested tables   11mm

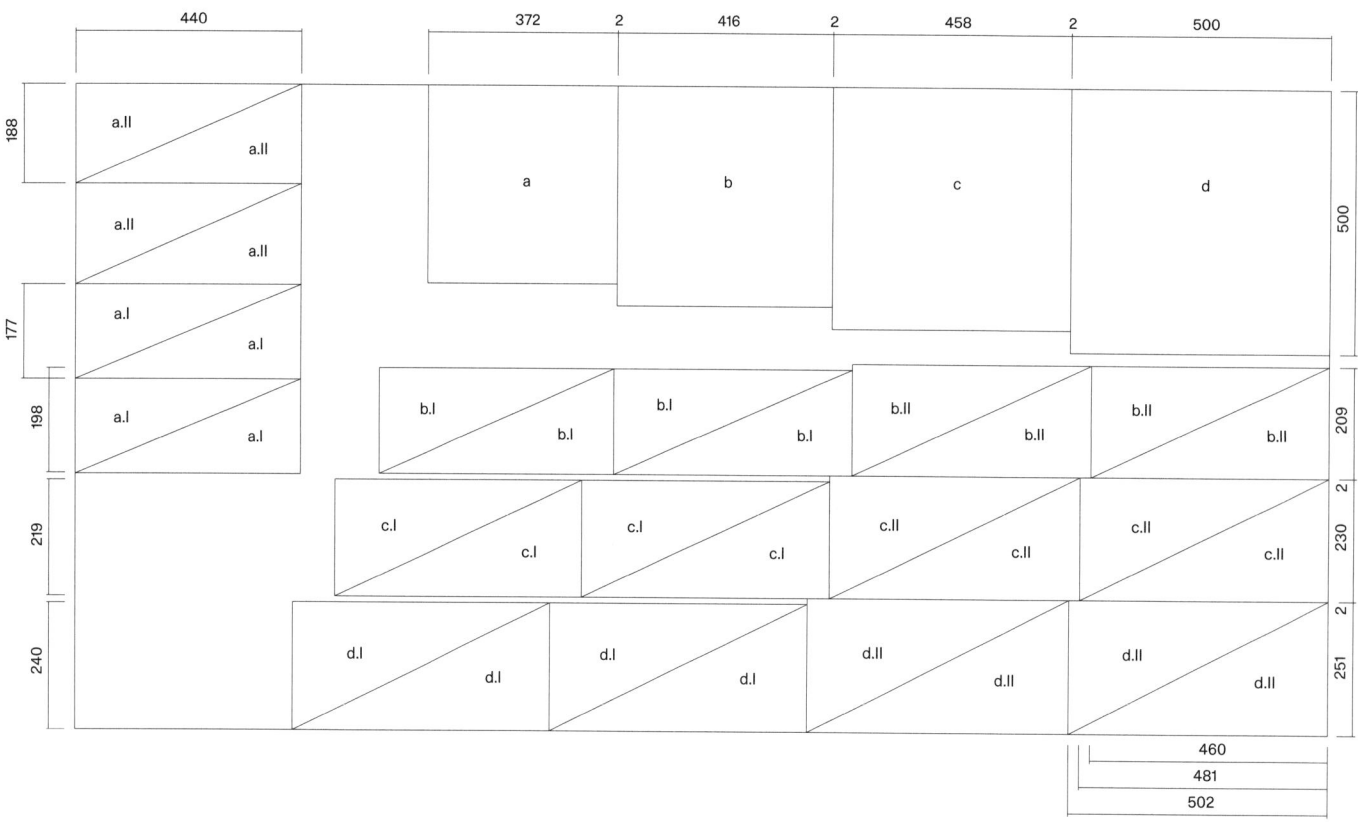

48

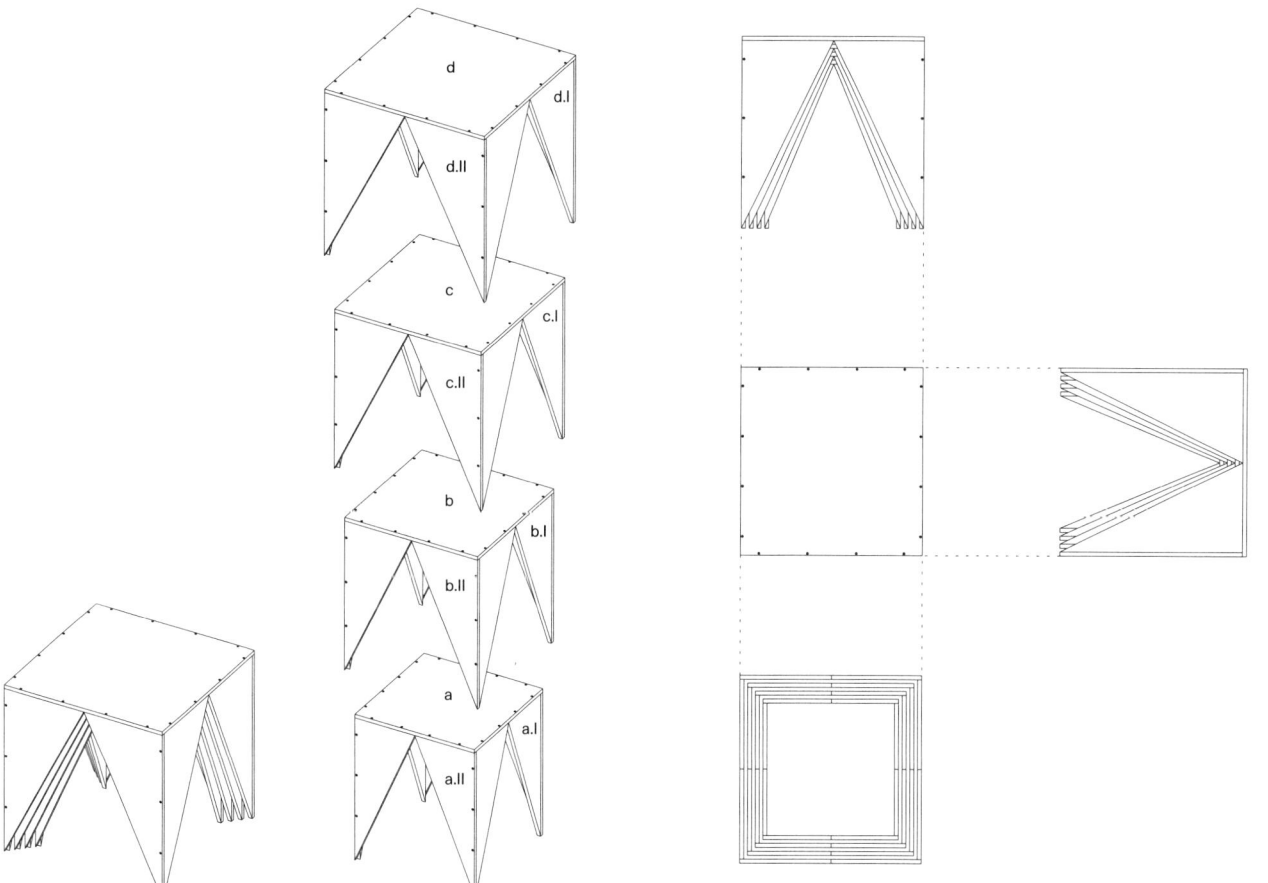

13.4 console 11mm

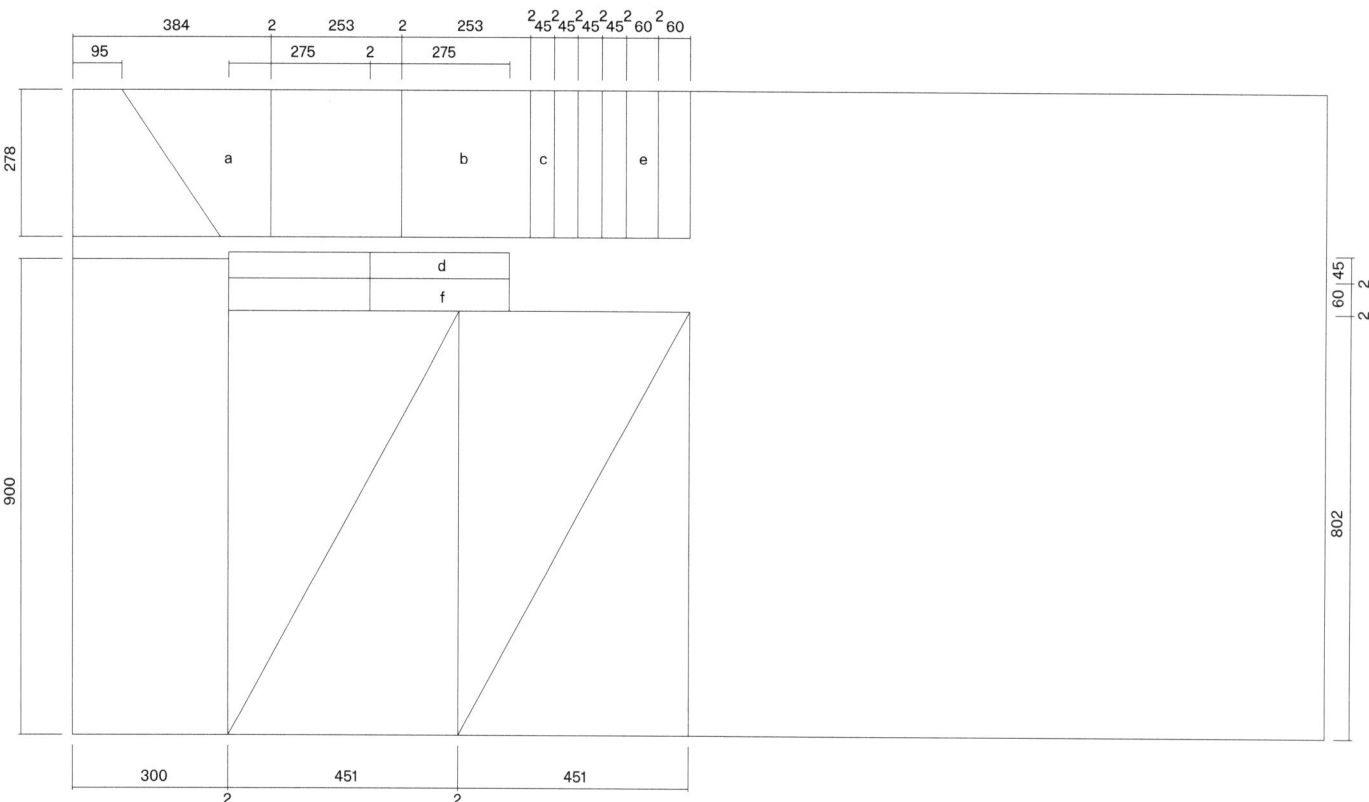

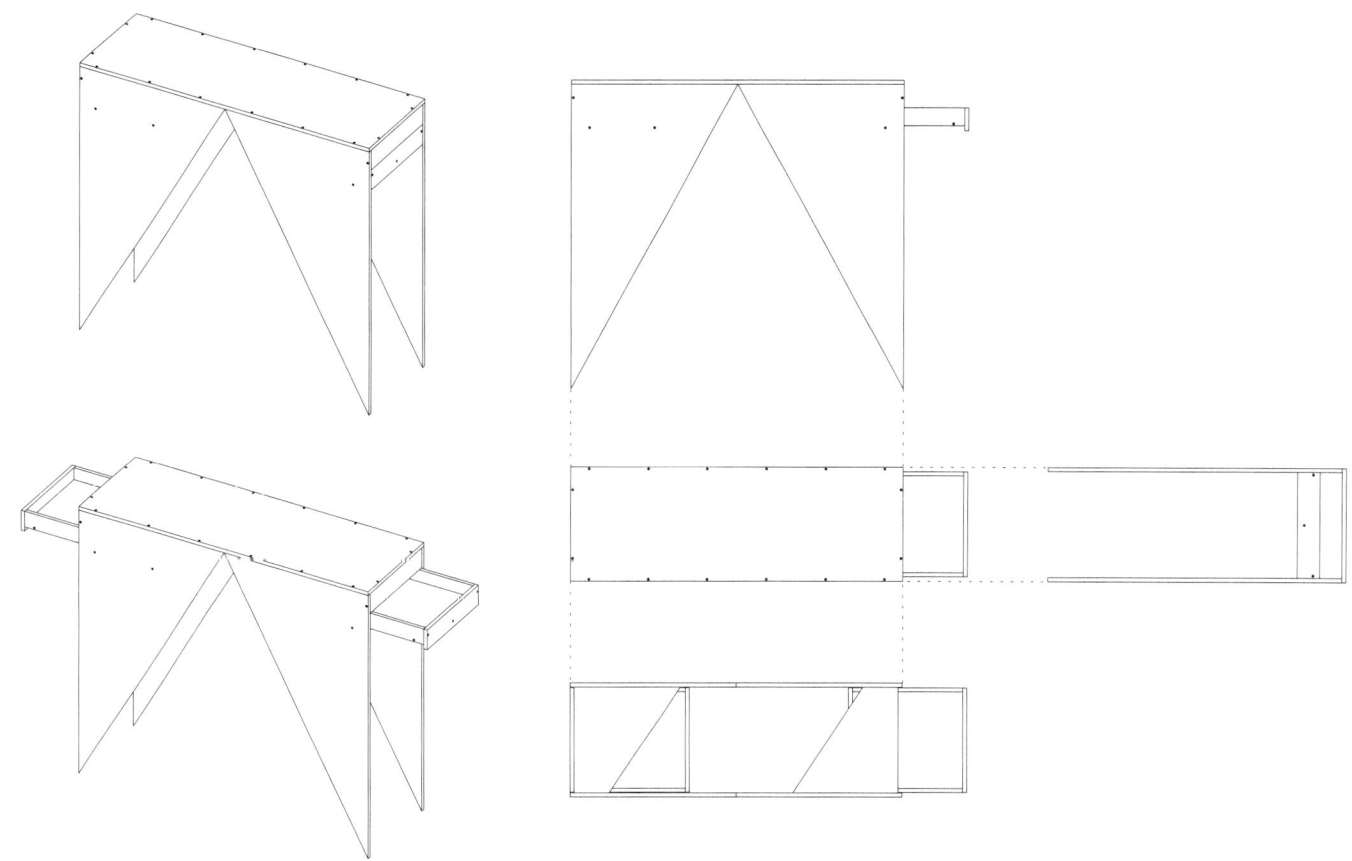

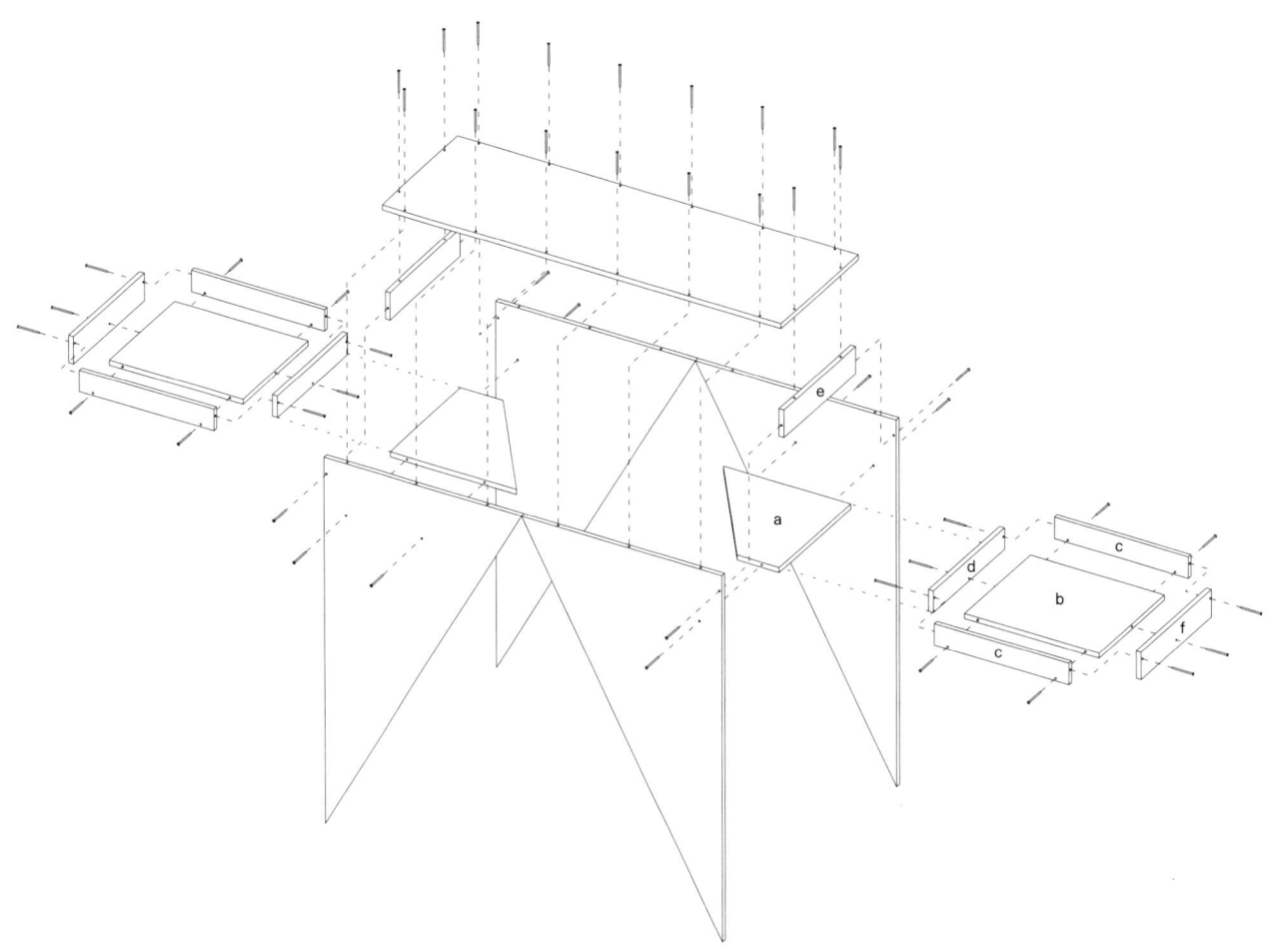

14.0 writing table   11mm

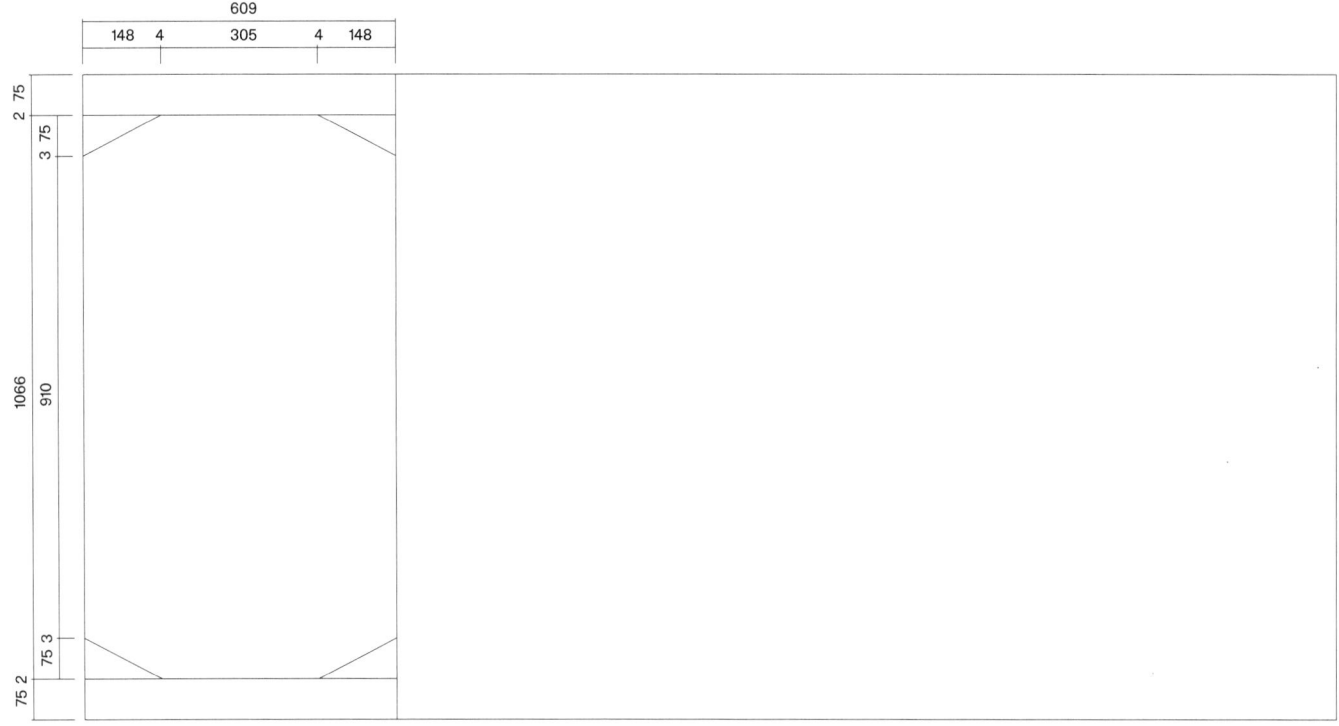

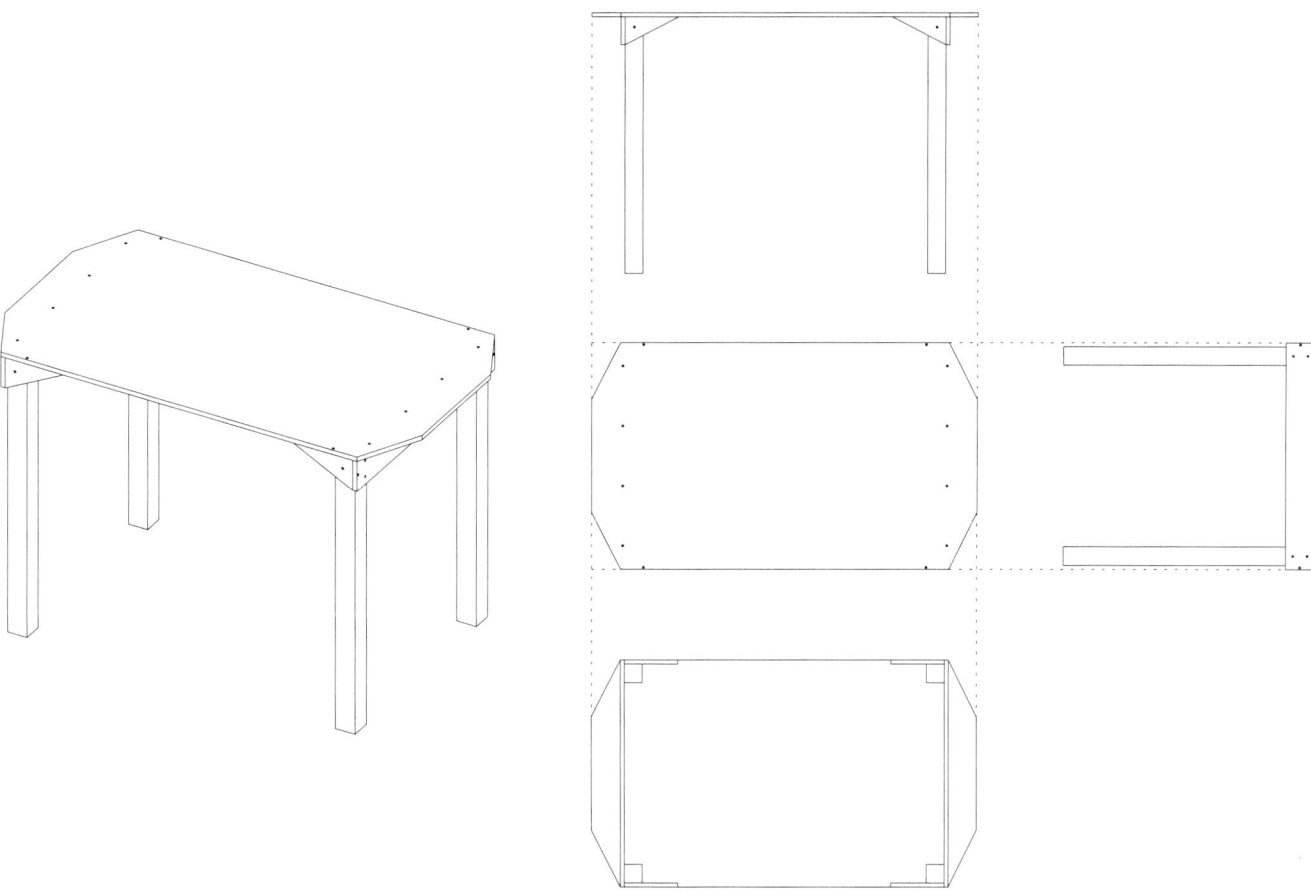

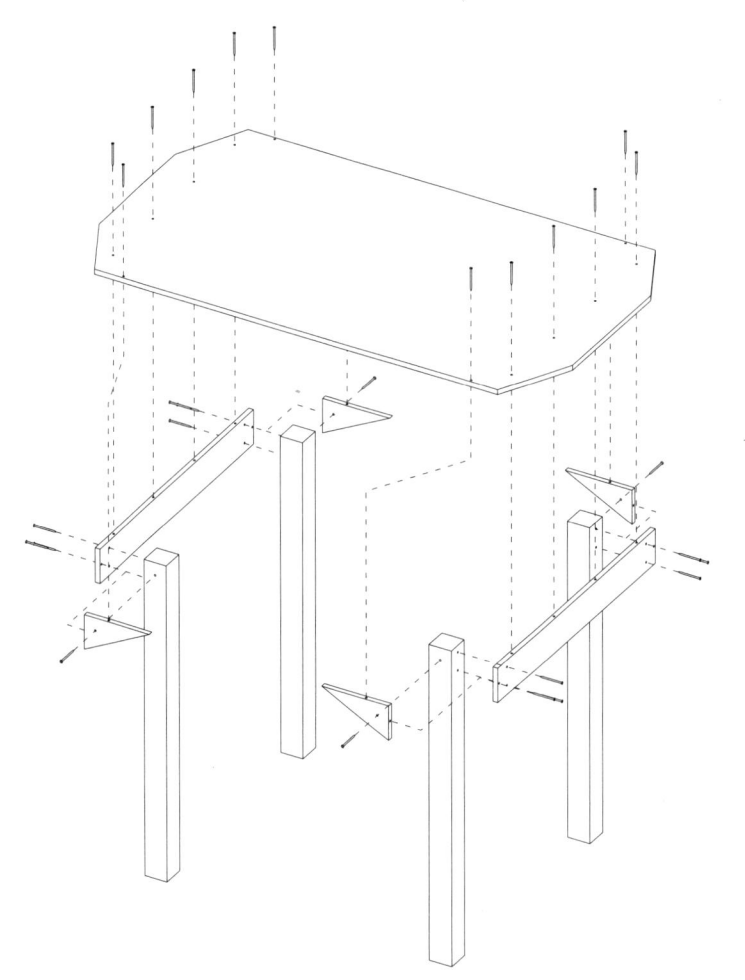

15.1 corner shelf 11mm

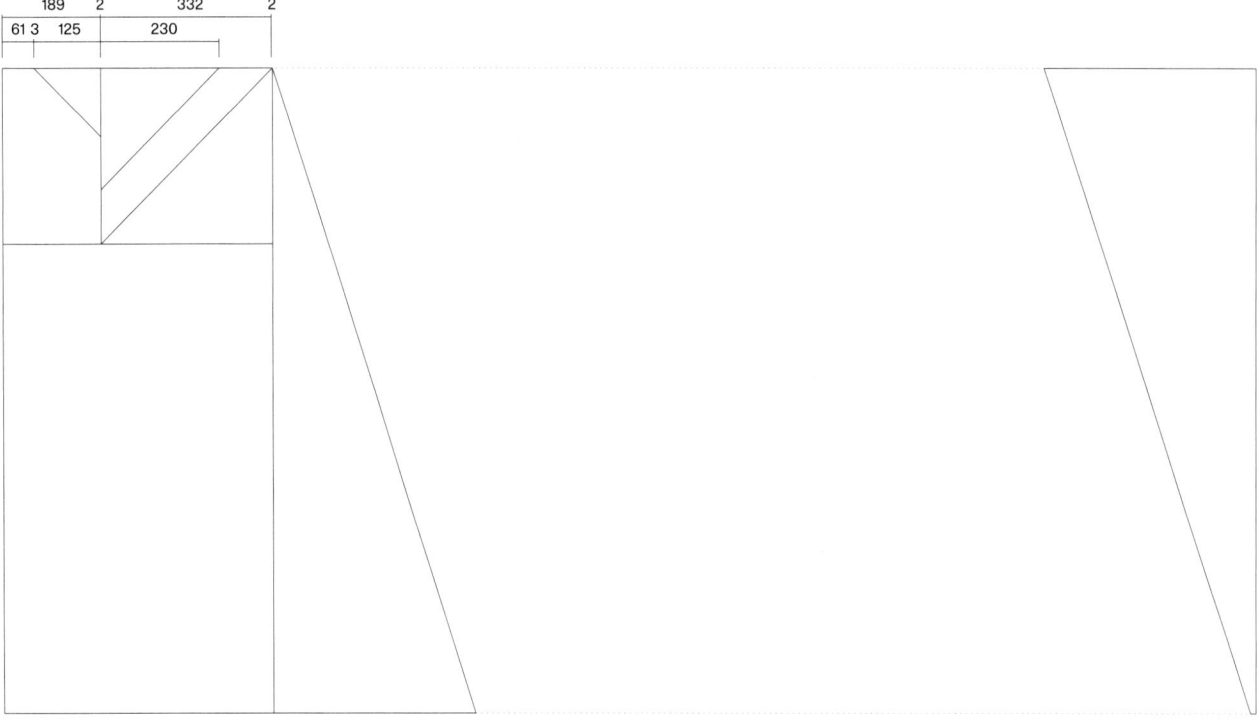

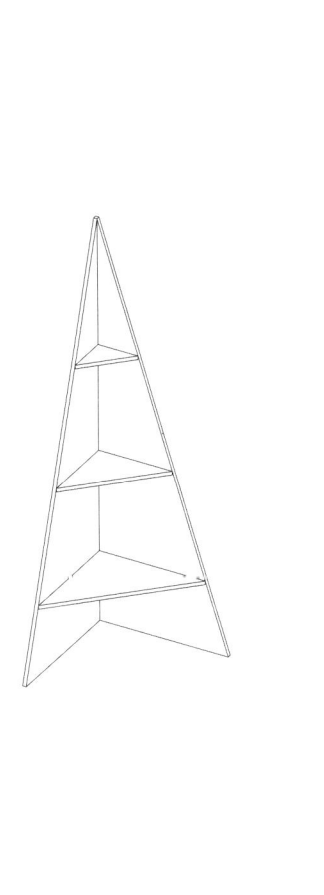

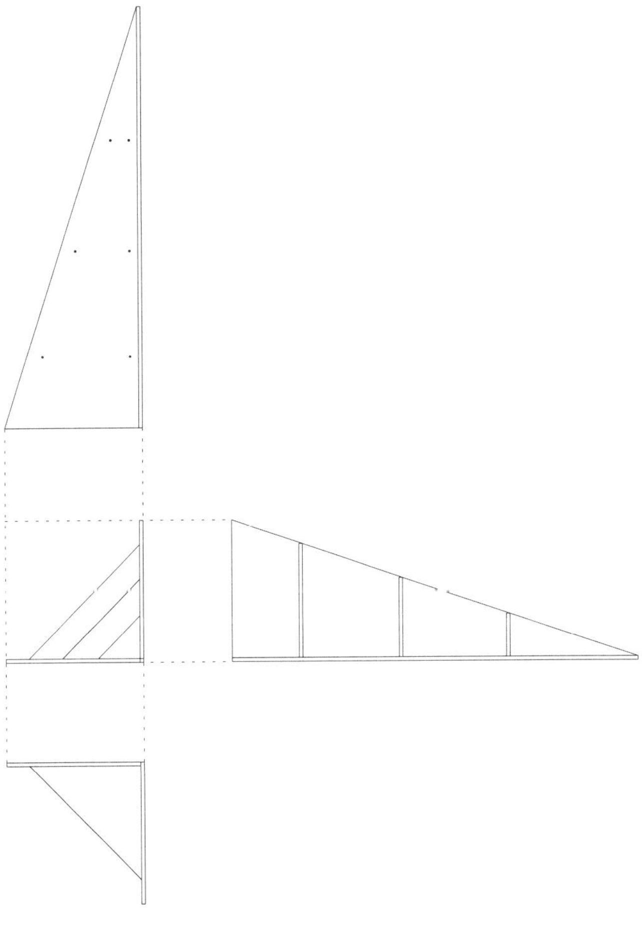

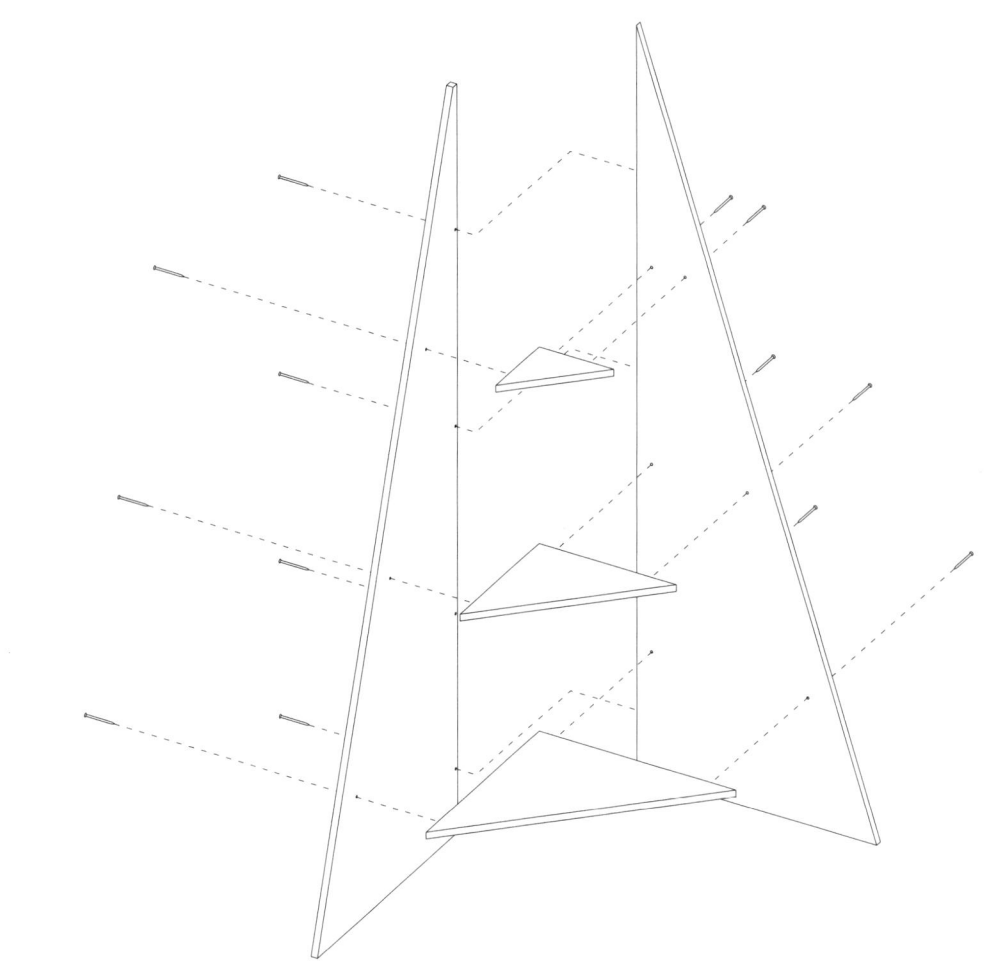

15.2 picture rail   11mm and 19mm

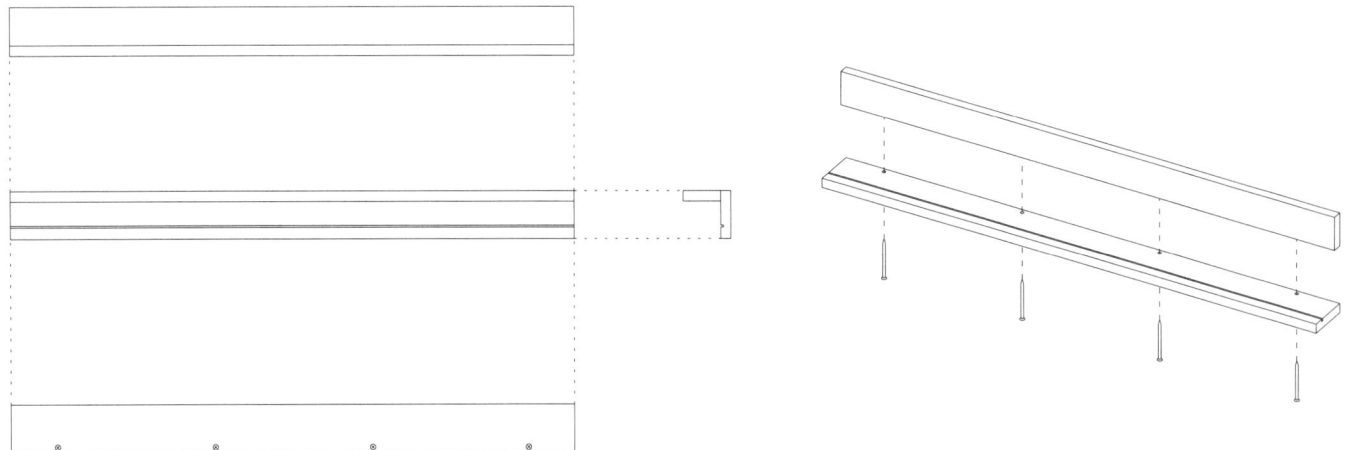

15.3 nail box 19mm

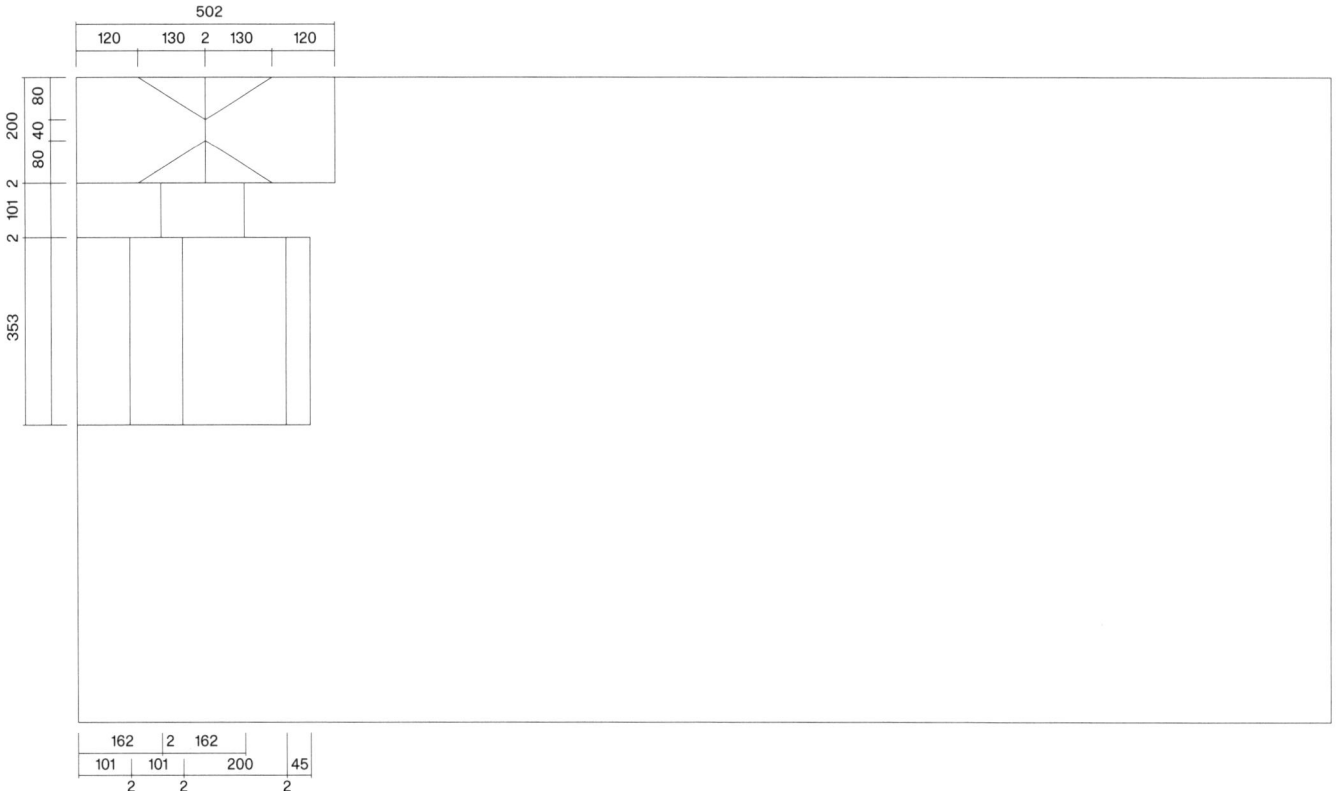

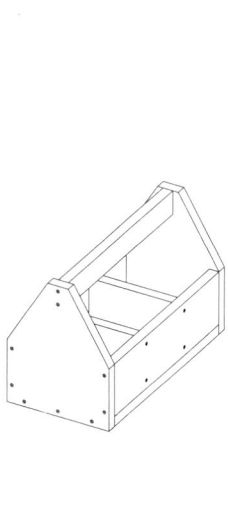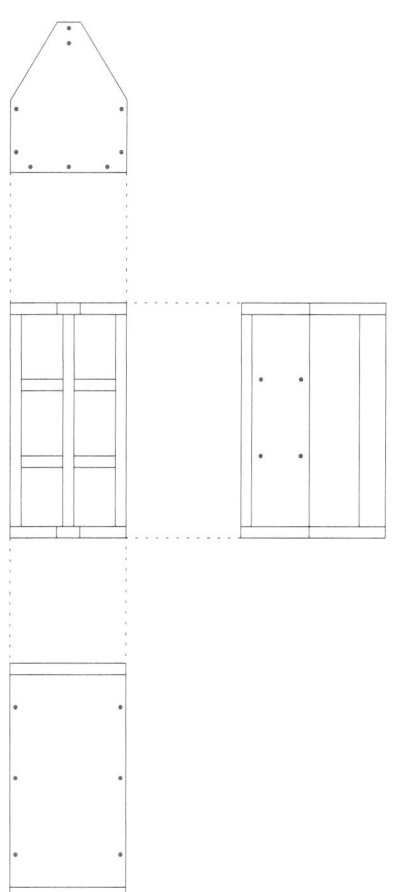

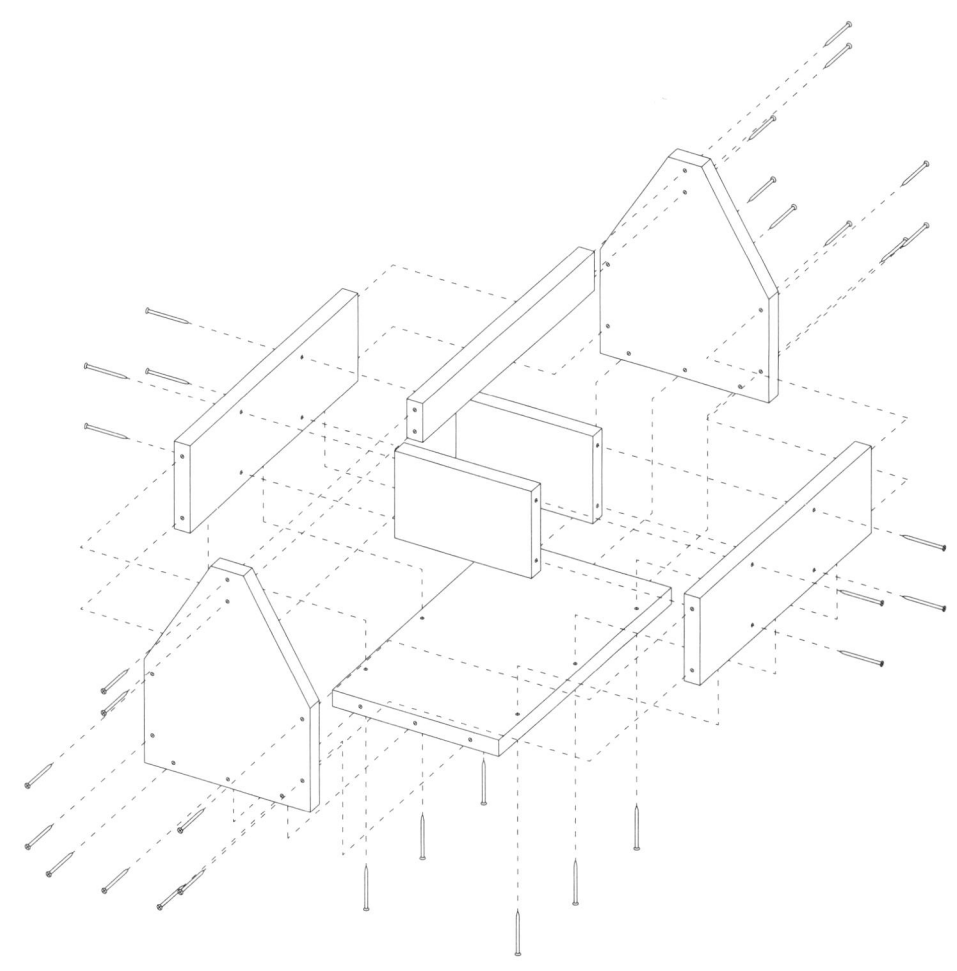

Set 1　11.2 coffee table ×1, 11.3 stool / side table ×2

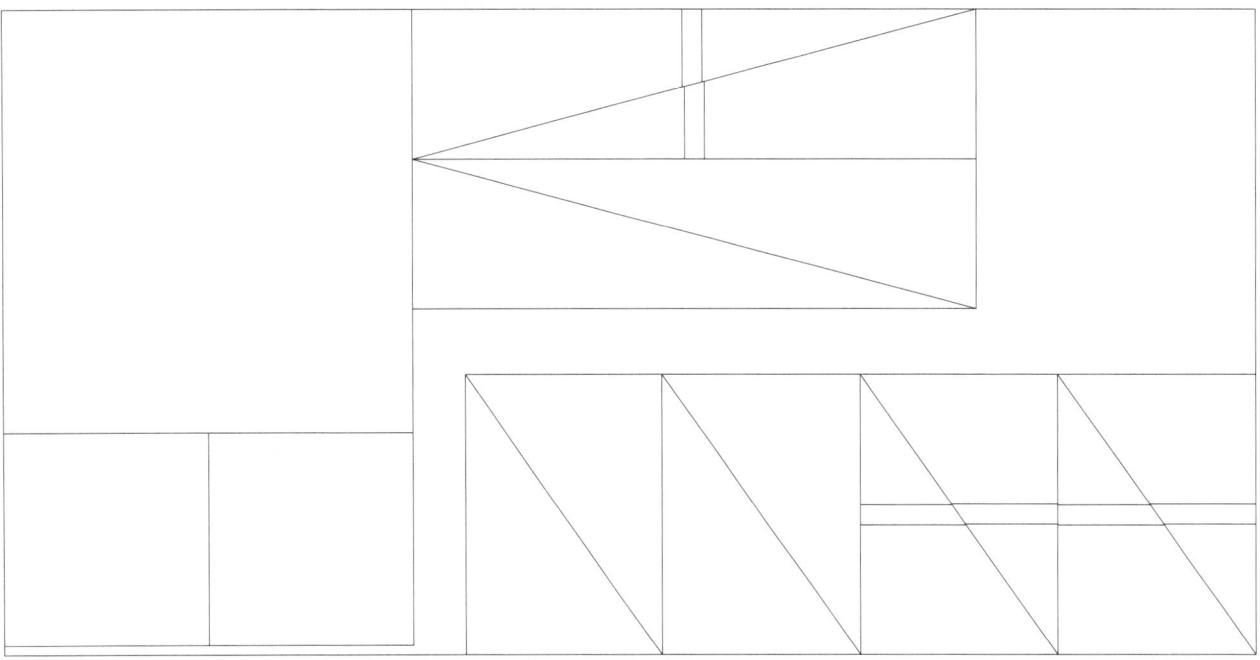

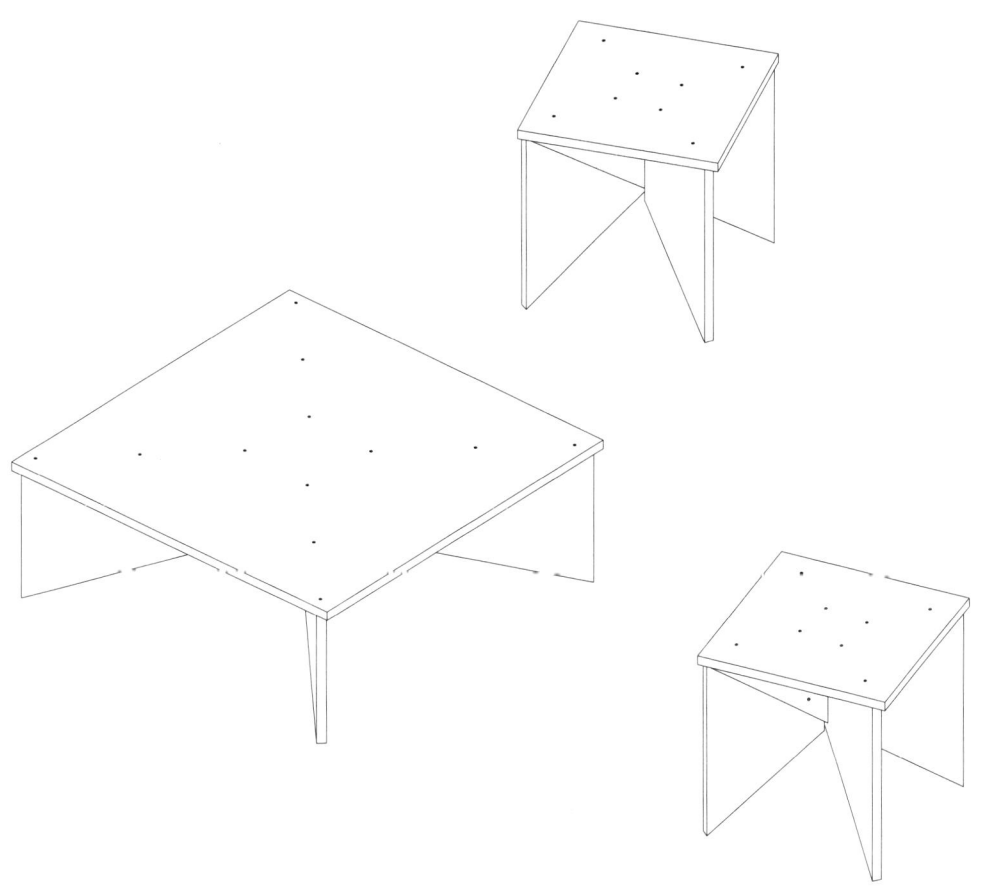

Set 2   11.4 coffee table ×1, 11.3 stool / side table ×3

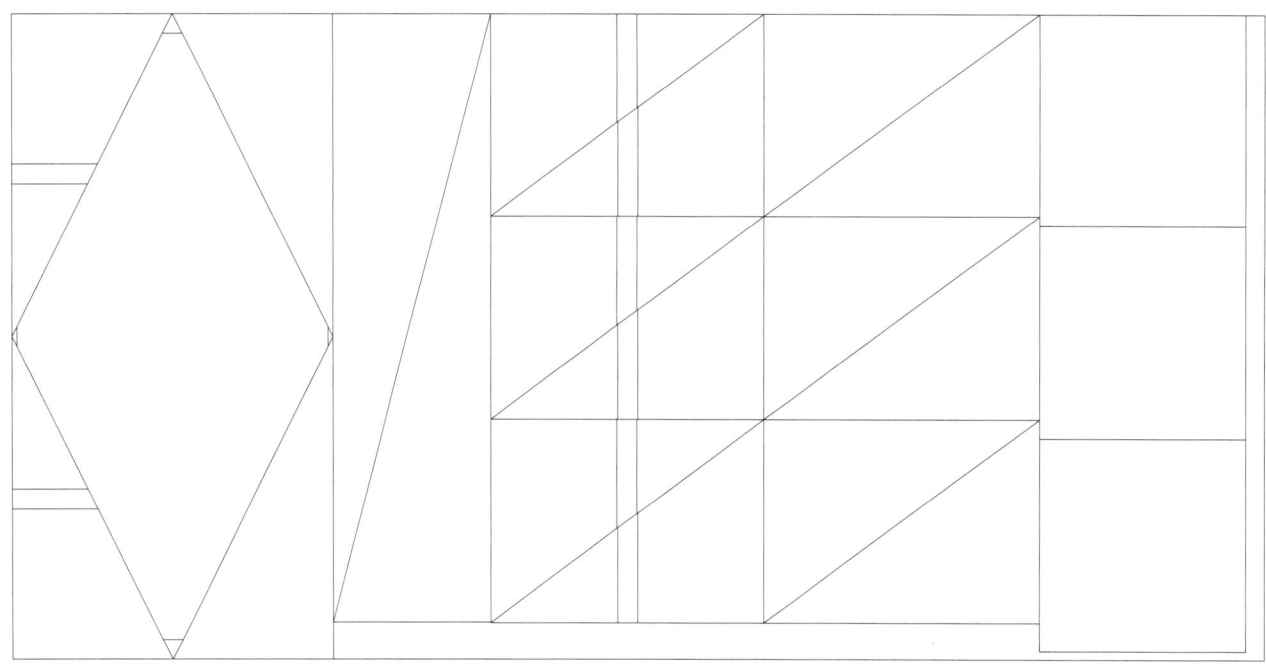

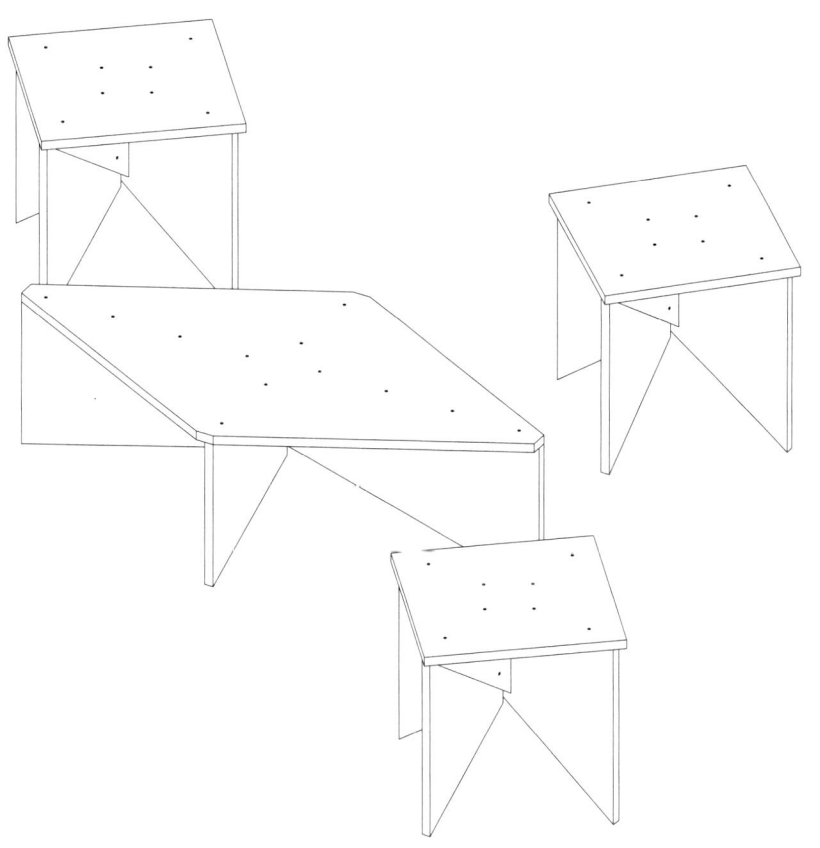

Set 3   13.4 console ×1, 13.1 coffee table ×2

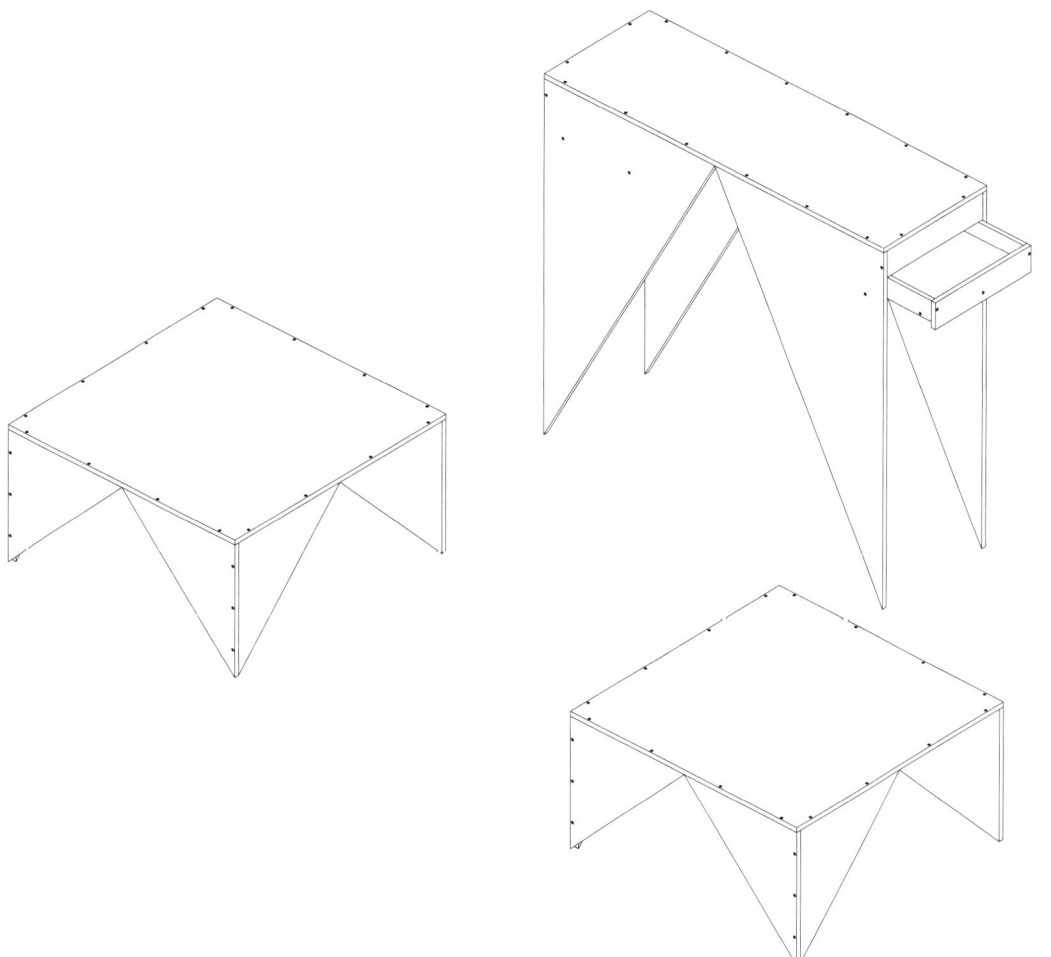

Set 4: 11.5 day bed ×1, 11.4 coffee table ×1, 11.3 stool / side table ×2

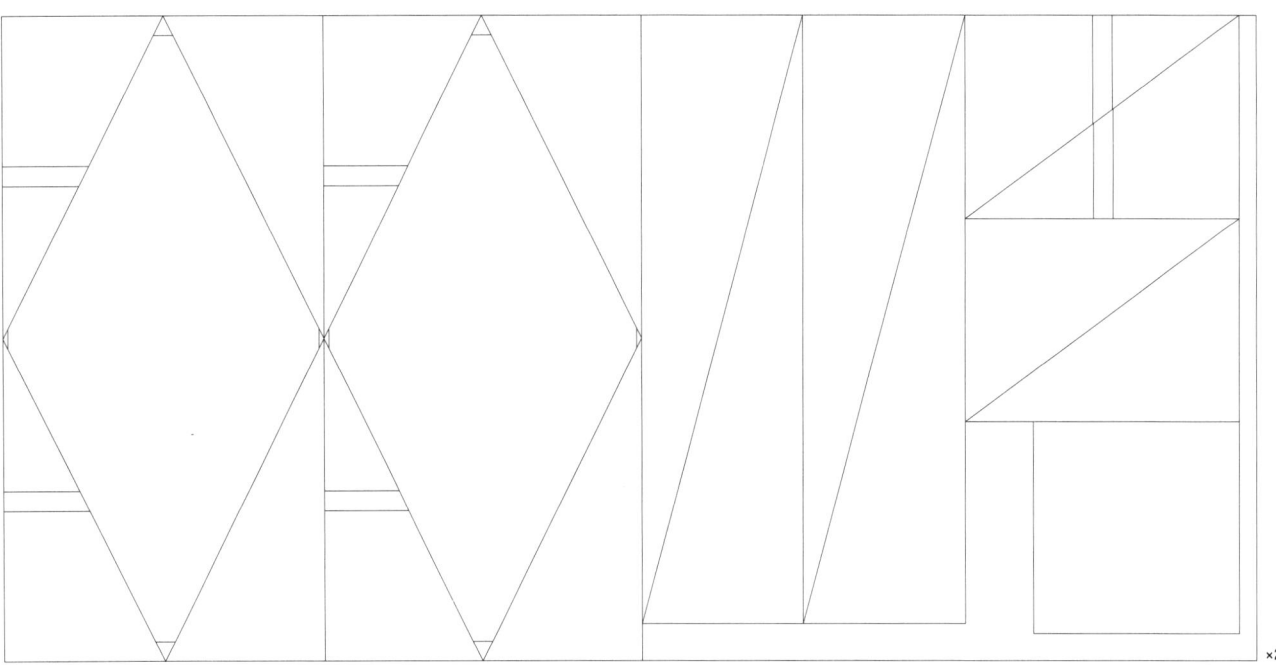

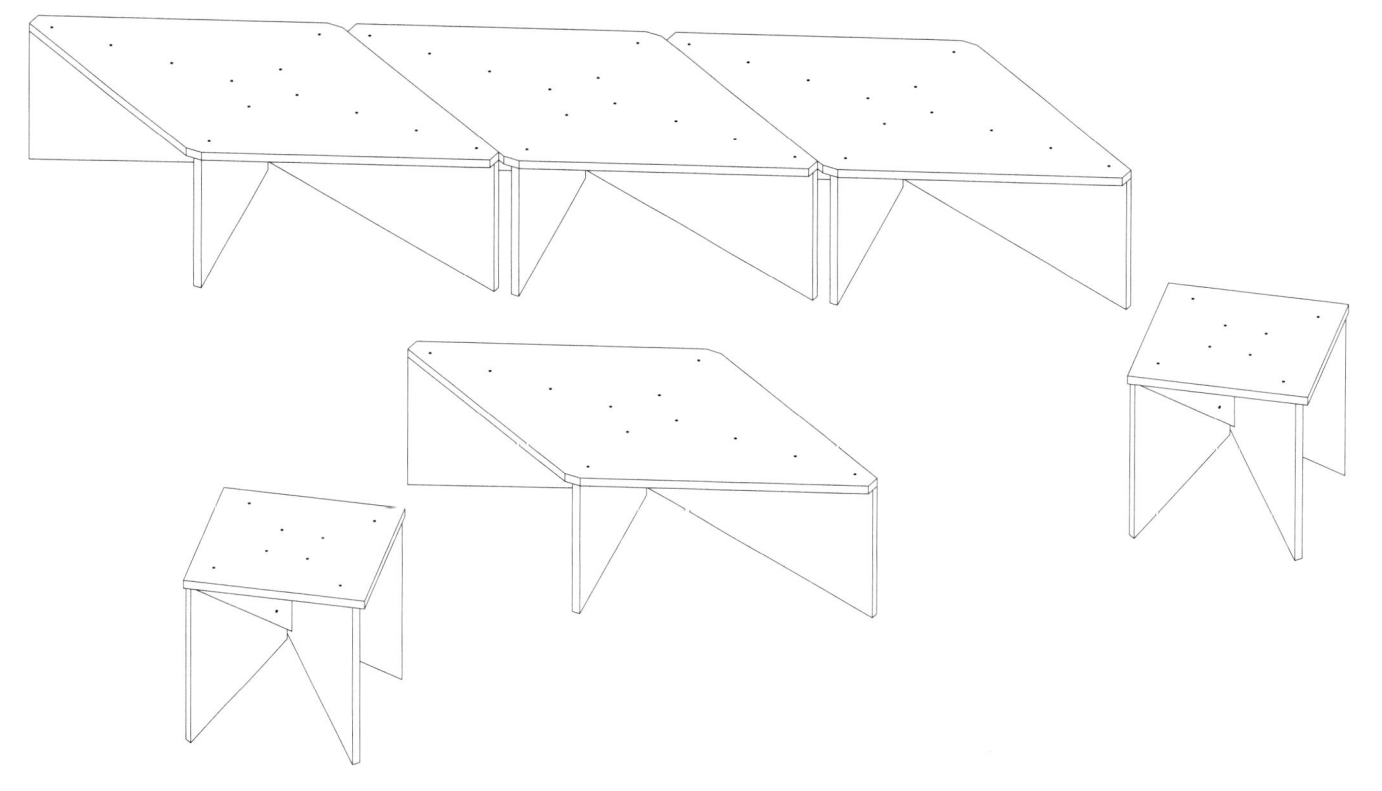

Tom Emerson and Stephanie Macdonald work together in London, where they live with their son Laurie. Their practice, 6a architects, is best known for contemporary art galleries, artists houses and studios, and education buildings. They have won several RIBA awards and the Schelling Architecture Medal in 2012. Their first book, Never Modern, 2013, charts the philosophical touchstones of the practice with critic Irénée Scalbert. Dust Free Friends, commissioned by Maniera, extends their thoughts to furniture.

Dust Free Friends by 6a architects

MANIERA
Limited edition furniture by architects & artists
74, rue de la Caserne
1000 Brussels
+ 32 494 787 290
maniera.be
info@maniera.be

© 2015 6a architects, London
6a.co.uk
dustfreefriends.co.uk

Designed by John Morgan studio
Printed in Luxembourg